ARTIFICIAL INTE

IN REAL ESTATE INVESTING

How Artificial Intelligence and Machine Learning technology will cause a transformation in real estate business, marketing and finance for everyone

DISCLAIMER

BOOK DESCRIPTION

Congrats on purchasing this book. You are about to delve deep into the future of real estate investing, property management, real estate financing and marketing. If you are excited about the industry or even just curious about the future of the real estate industry, then this book is for you.

The book starts off with the history of Real Estate Investing in the United States. From the days of early colonization to modern use of online rental applications, it goes through every stage in the process; so, you know how we got to where we are today.

Then the book talk about how we can use advanced artificial intelligence tools like machine learning and neural networks to evaluate property prices; in addition to using tools like VR (virtual reality) to improve the customer experience.

The book then talks about how the role of real estate agents will change with introduction of high efficiency algorithms to filter applications, complete paperwork and properly evaluate property rates. There will also a significant increase in security with facial recognition systems and reduction in maintenance costs with use of home sensors.

While the real estate industry has been slow in adapting digital marketing, tools compared to other industries, research indicates that use of these tools will significantly reduce customer acquisition costs. It's just a matter of time before it is used all across the real estate industry.

Then, the book looks at how artificial intelligence will significantly improve the way landlords pick the best rental applicant. While credit score and current income are reasonably good factors; they miss out on a lot of critical factors that artificial intelligence can easily capture.

TABLE OF CONTENTS

11

THE HISTORY OF REAL ESTATE INVESTING TECHNOLOGY

Early History of US Real Estate

Real Estate Evolved Differently Depending on Region

Real estate in the United States was very different during the founding days of the nation. In fact, most of the land that was sold was sold in huge quantities. Most people were farming back then as that was the only way to truly live.

Essentially, settlers would first come to the land and just build a house there. Over time, more and more of the land was owned by settlers. The safer regions were owned by someone else a good portion of the time.

The solution was to sell part of the land to the newcomers or to allow children to inherit parts of the land. For a very long time, new houses were not often built because the local lifespan of the individual meant that the child what often inherit the house.

However, this presented a unique situation where real estate laws evolved based on areas that were settled first. Laws around selling homes in Pennsylvania was different than laws about selling homes in Missouri because they were settled at different times.

Rapid Evolution Brings Historical Difficulty

This uniqueness is intriguing which means that pretty much anything was possible. It is very difficult to historically define the early beginnings of real estate beyond the general sense. The laws of the past have been changed and edited for the future with little to no record keeping.

However, it does present a unique case. Real estate evolved as more people entered an area. This is true even to this day. Real estate evolves every time that the previous rules can no longer adequately serve the new area's population. Just as we originally started out by just setting up houses wherever we stopped, we also started selling the chunks of our land to whoever might happen to be interested.

Eventually, we began to face scrutiny by the government. The colonies, as American cities and areas were first referred to, faced higher taxes on property and items than the country of origin. This eventually, along with other factors, led to the revolutionary war.

The Government Began a Mass Land Collection

What started to happen was the people in an area began to form their own laws to keep them safe. These people would enact local governments that would represent them. These local governments would then employ people to keep the security of the land.

Local governments were very small in the beginning. However, when the government begins enacting services, those services needed payment. The services are provided by people who need to live, thus they need a reason to do the job. The government didn't really have many options to bring in money. Being able to collect land, on the other hand, provided them with a way to collect money. In addition to this, sales tax and other taxes persisted in the original 13 colonies. Sheriffs in outward communities were often paid by bigger cities that sent them there.

Banks

Banks Acquired Quickly

Banks were relatively localized and just as real estate evolved with the area, so too did the practice of protecting money. People needed protection from outlaws when it came to their money, which meant banks had an opportunity. A bank back then only meant that your money was in a safe. This is where the old cowboy trope of breaking into safes comes in.

However, banks had to make money somehow and they couldn't just take their customers money' although some did and still do. The primary way that Banks made money in the past was by the owning and selling of property. Essentially, some lone farmer would fail to get a family. That farmer would die and the town would designate the new residence to the bank.

The bank would then go through the process of selling the house to those that needed it. After all, if some person dies they can't sell you a house. Banks would often play finders keepers in such a situation. This allowed Banks to grow in more populated cities and eventually they were known for selling houses and holding money. Specifically, they held deeds and money.

Deeds Were A Thing

It is very rare for the average person to see a real-life deed. There are a few spots of land in America that still have their deeds held by private citizens. A deed represented that you fully owned the land and that you didn't owe anybody anything. In fact, income tax and things of this nature weren't around back then. This is because there was no official government that

18

needed funding except in the colonies. However, those banks often held land they made money from.

As the government got bigger and we progressed through the Civil War, the government started collecting the deeds. Essentially, the government became a huge bank and started obtaining land. They would get the deeds of people who died and then get the deeds of people who were in financial trouble. Essentially, they started targeting the lesser-known people before they went after the bigger people. Eventually, the class of rich tended to cleanse itself, with the government obtaining huge amounts of land during the Civil War. Civil War represented the national government trying to get control of state governments.

Those who participated and lost or died on the wrong side in the Civil War often found themselves losing their land. This would be referred to as the spoils of war. This allowed the government to amass a huge amount of land very quickly by simply winning the war. There are still some spots in America that have deeds attached to them but most of the deeds that are in existence for America are held by the government. These deeds represent claims that the government owns the land, which allows them to operate on the land.

A deed represented a document that said that you were kind of your own nation. By having the government take the deeds of nearly everyone, the government wrested control from the people. This allowed banks to rent houses and land to people as a form of revenue rather than solely relying on selling houses and growing interest on money.

Banks Acted as a Local Real Estate Agent

However, that is a lot of property to manage and the Government had a lot of trouble with communication and management. This was still during an era where the phone and electronics were new. In fact, there weren't many institutions dedicated to something like the news. Such news magazines were only seen as viable items for the rich or the almost rich.

This meant that the government had to find a way to sell land and houses so that it could begin getting money. Therefore, Banks primarily became the representatives of the government when it came to selling and maintaining the land. Banks had already spread through most of the United States and represented areas of control. This would allow the government to sell the land it needed and Banks could join in on the money to be made. Therefore, the first local real estate agent could really be considered to be the banks of America.

Bankruptcy, Crimes, and Real Estate Companies

Bankruptcy and Rundown Homes

While it was possible to obtain run-down homes in the past and obtain homes through bankruptcy, it wasn't prevalent. Once Banks started acting as real estate agents and landlords for the government, it became prevalent. The standard process was that if you couldn't pay land taxes, the land would be taken away from you even if you could pay property taxes. Essentially, it was a very easy way to justify taking land away from families who had been there for generations.

These events made it possible for the selling of land to become more commonplace. Run-down homes were usually left alone because it would be miles before you saw anyone else. Therefore, it might be years before somebody managed to find your house and it was almost always assumed that someone owns the house. Therefore, if someone died in that house then no one really knew about it for a long time.

What would happen is if you missed a monthly payment then the bank would investigate. This shortened the amount of time that property would be found and sold. Either you were bankrupt or you were dead. In either case, the bank could go about selling the property once more and generate even more revenue.

Crimes and Auctions

However, this didn't really solve all the problems because you still had criminals that took people's houses. If a bank obtained a house that was previously owned by a criminal, they didn't want to bother with any of the repairs. When we refer to a run-down home, we're talking about places with missing roofs, windows, and other items. In modern day terms, it would be a house that didn't have water or electrical wiring. It was a massive pain to try to maintain these houses. To save on cost, they simply sold the land for less hoping the owners would create a good house.

This led to a small market of people that would buy houses that were not up to date or in good condition. These people would then begin preparing them and selling them, much like the way we do it now. However, buying and fixing them usually didn't take as long as it does now. There are a lot more standards that one must follow when repairing run-down houses such as Florida requiring water. Therefore, one, back then, would expect to buy a house and be able to sell it by the end of the month. This created a rapid expansion in real estate.

Where There's a Will There's a Way

The sale of houses was primarily designated for the banks and the individual person still. There was a small collective of people that were individually selling houses as a form of extra income. However, due to the price of houses, this was primarily a practice of the rich. The rich were able to buy lands outright, build smaller housing to fit the usually large land, and sell individual houses. It was the rich that was able to divide the land into smaller housing, which is why you often see very cramped big cities. These cities weren't cramped in the beginning, but because of the real estate market, they became that way.

The price of a house went down significantly with a lack of land. You could buy a plot of land that could fit 10 houses on it, but the incentive of the land was usually that there was land. Therefore, if you were selling each house for $10 you would have expected to buy the land for maybe $50. Thus, buyers thought they were saving money when they were given the house for cheap (without the land). But sellers made a lot of money on the exchange. Buyers thought they were buying a significantly cheaper house in exchange for a loss of land.

This is because building these houses usually required quite a bit of personal labor or hired hands. If you were a cashier, you are not likely built with the ability to make a house. In fact, one could say it be quite dangerous to do so. Instead, people began specializing in building houses and they would be hired to do so. They built houses fast and with the cheapest materials they could find.

This is unlike the materials found in homes prior to this area. This is because homes prior to this era were built by the individual so they were often quite sturdy and they were overbuilt in most cases. These building companies would use math and building knowledge to lower the cost, time, and materials needed

25

to construct the building. These are the people responsible for creating the numerous health hazards that come with older buildings.

The average person didn't know about this type of business unless they dealt with it directly because it wasn't widespread for quite a while. Many of the cities in the 13 colonies were the first adopters of such practices. This is because these areas had the largest population.

The Real Estate Era Began

Eventually, the cities got bigger and the populations became unwieldy for these individual sellers. Therefore, since multiple sellers got a reputation, people wanted to learn from them so they could earn more money. This led to rich individuals hiring friends or family at first and strangers later to participate. These were the first real estate companies as many employees were located in one building.

Running it as a business afforded these companies an additional opportunity where some skills were used more than others. Some of the employees were focused on solely finding houses. Some of the employees were focused solely on selling houses. And some of the employees were solely focused on the paperwork involved. In other words, by having multiple employees for the Real Estate Company, these companies could sell far more than they could all do individually.

These companies expanded to begin selling in different states and that is pretty much the story from there for companies. These companies have been around for a very long time and it's not going to stop soon. However, while companies have existed, that doesn't delineate the fact that the individual person can also still sell houses.

The Individual Agent

The Individual Sale Never Went Away

The truth of the matter is that the individual sale never truly left. The only benefit that companies had was that you could sell more at a faster rate. However, this is only achieved when people are selectively tasked. These companies designated jobs that people were faster at. Therefore, the work was divided. A person's work was divided three ways. One for looking, one for filing, and one for selling.

This primarily was used to save time because by dividing it into three, one-third of the overall time was spent. This gave companies an advantage over the individual meaning they could make more money. However, that didn't stop people from trying to sell their own houses. There was plenty of real estate on the market for individuals to sell houses, but companies had a leg up on them.

When a person goes to sell a house, they want to get the most out of their deal. This means that they usually choose a company because there is more safety. It's much more likely that a company is going to be able to sell a house for the maximum value than an individual. This may not be true, but this is generally what the perception of a company versus the individual is.

Companies might have the power to get all the houses from individual sellers. There are houses that companies don't want to sell for various reasons. They may not have real estate agents in that area. That house might not make them a lot of money. That area is not favorable to selling houses such as high crime rate or low-quality schools. There are many reasons why companies don't take all of the houses on the market. However, the individual didn't really have access to these houses until something was made.

The Classified Ad

Most transactions either happened as a company or as a close neighbor. In other words, if you didn't live near the person you wanted to take part in selling the house, you couldn't sell houses. It was not a very convenient business to venture in. Companies had the ability to push their name out to people whereas the average individual did not. This meant that getting a hold of houses as an individual was incredibly difficult.

The invention of the classified ad was born because people wanted to sell items their neighbors didn't want to buy and companies didn't want. The classified ad was an exchange section, like the old version of Craigslist. It was an area of the newspaper where one could find work, furniture, and bargains. This allowed the average person to go about selling houses and it was also word-of-mouth.

You see, even though the classified ad section made it possible to find houses, it was often the place where you would see individual sellers. This would be the section where small-time real estate agents would sell houses on behalf of other people. These other people would hire them based off family and friend connections. Additionally, these real estate agents would go out and try their best to entice other people. What ended up happening was that the individual seller had a market, but it was small. This led to people selling houses in their free time rather than relying at on it as a job in it of itself.

The Phone Line and Adverts

Around this time, the phone line became a thing. The phone line interconnected cities together. What this meant was that cities could now more effectively sell to each other. If you were in New York and you wanted to buy a house in Philadelphia, you would look for adverts from Philadelphia. This is because people in Philadelphia would call to New York to place an advertisement for a house in the Philadelphia section of classified ads.

More specifically, ad agencies would be hired to place ads in several cities for a certain cost. An individual real estate person would hire the temporary services of an ad agency to push their ad to other cities. This would only happen for the rather expensive houses.

Companies, of course, took up the bulk of advertising to other cities but they had bigger advertisements. Normally, instead of putting a house on the classified ads, they would put a bigger advertisement with their number on it. This would lead people to call them and see about available houses. The companies would then call each other in their respective cities so that there was a company-wide list of available houses.

The individual person, on the other hand, would put their version of this in the classified ad. They would both put their personal advertisement for selling homes and the homes that they were selling in the classified ads. This allowed them to reach a wider audience to both buy and sell.

The Internet

The invention of the internet did a couple of things for real estate agents. First, it made the process of filling out applications far easier. However, we'll get into that in a little bit. The most important thing that it did is that you could hide that you were an individual seller and act as a company. The internet doesn't allow you to see the staff behind the website, which means you can act like you are a company.

What this did is it allowed individuals to have a false addition to their credibility. The internet also hasn't been very good for real estate agents because of fraud. If you can act like anyone on the internet, you'll have some people acting like you. In other words, a potential customer that's ignorant of how real estate works could hand over their house. It's rare that this might happen but that is the extreme of what can happen to a potential customer.

The internet certainly has brought many problems with many benefits. The internet has brought about faster ways of filing, documentation, and billing. It's easy to keep in contact with clients; and easier to market to potential clients. Generally, everything is faster. It is thanks to the internet that the individual can outsell companies now, but it takes a lot of dedication to do so. In fact, most of these individuals end up becoming their own company. Therefore, you have started to see a few companies pop online. So, let's go ahead and go over the benefits and reasons why the internet is preferable.

Online Applications

Buy Before the Move

The most beneficial side of selling houses online for the customer is the ability to buy before they move. Most of the time, customers buying houses in other cities are doing so for jobs. In the past, you had different stages for this. During the beginning, you simply moved to a new location and built another house. In about the middle, you would call a company and see if there was a house available. Nowadays, you simply fill out an application and start receiving emails about potential houses.

As you can see, it got progressively easier to locate houses in other cities. You don't even need to fill out an application most of the time nowadays. Instead, you could look up the price of the house along with the city and find quite a few listings. This has been the most beneficial side of bringing real estate online.

It reduces the amount of work on both sides of the transaction. The customer no longer must call daily to see if there are any new houses. The real estate agent doesn't need to bother the customer every time there's a new house. Ultimately, this saves time for both the customer and the agent.

Google Maps

On the customer side, they no longer actually must go to the house. They can simply look up a house in Google Maps and go through the pictures attached to the address. They can view the neighborhood from every angle possible. They can even inspect the local businesses to see what's going to be convenient.

The only problem with Google Maps is that it usually has an update time of about six months. It's always going to be past information. Therefore, the addition of Real Estate websites has included personal photos. These photos inform the consumer what the house generally looks like.

Not only this, but you don't have to worry about getting lost when going to the house. In the old days, you pretty much just pick the spot at the city you wanted to be at or county. Then, during the ages of phone calling, you were giving directions that weren't always accurate. You had to specifically rely on that company to tell you how to get to that house once you were in the city. You really didn't have any idea beyond road signs about how to get to that City. Also, phones were landline so if you wanted to talk to the company again, you had to borrow someone else's until Phone Booths became a thing.

Nowadays, you type in an address in Google maps and you get a full path. Not only that, but you can also select a path that doesn't have any tolls. You can select the path that's not on the highway. It is nearly full of customization options to help you get to your location in a way that you prefer.

This means that if you are traveling to another job and using an RV in the meantime, there's an RV route. If you are a Semi driver, there's a Semi driver route. Google Maps saves a lot of time for consumers and prevents real estate agent time from being wasted on just giving directions.

Company Cooperation

Over time, companies began to provide tools to other companies. You have companies like Office 360 to handle documentation. Companies like QuickBooks to handle billing. Companies like PayPal to handle invoicing. Essentially, you have tons of tools that simplify and expedite Services previously paid for.

If you are a single real estate agent, documentation, billing, and invoicing is not complicated. Unlike a company where you might have multiple real estate agents, you are one person. You don't charge that many customers in a single day. You have about the same bills you would have in just owning a house. Lastly, your documentation is relevant to how many clients you have at that time. It's a rather small workload compared to what it used to be.

All of these Services work in cooperation with each other. QuickBooks works with bank accounts and things like PayPal. Office 360 works with things like QuickBooks. These companies work together through programming to create an ecosystem. This ecosystem helps you be a business as an individual. By doing so, you benefit from not only company cooperation but also the reduced workforce. Yes, you don't get as many clients as big companies, but you do not have as much work overhead.

AI and VR

Reduction of Driving Time

VR is a fantastic addition to the real estate space. You no longer have to worry about meeting clients at houses if you can do it in VR. The way the VR works in real estate is between two different devices. You have a house scanner, which you bring into a house and it scans the house itself. This scan produces a 3D model that can then be used in a virtual reality environment. You then use Virtual Reality tools like the Oculus Rift and the HTC Vive to look at the house as a 3D model.

This investment is extremely low cost because you only need to buy a computer that's powerful enough to run VR. There is a software company that is currently the only known one on the market that provides scanning. The scanner is extremely expensive to do it yourself, but you can hire someone to do this for you. These scans usually run a couple of hundred dollars. However, you can then reuse the 3D model however much you want and never have to force a meeting with your clients.

The reason why this is a fantastic addition is that not only do you cut down on driving time, but you also cut down on viewing time. Normally, real estate agents give their clients about an hour to a half an hour to look at a house. There are only so many hours in a day. With VR, as many VR systems as you have you will be able to show off a house. That means if you have 12 clients, normally a full day of viewing, with six machines you could be done in 2 hours. This saves a lot of time, time that can be spent on getting and selling more houses.

AI: Worth and Management

AI has done a lot of things since its conception. In the world of Real Estate, artificial intelligence can do a number of things. While we will get into this deeper a little later on, artificial intelligence can tell you the worth of your house and a potential selling price. In the past, you would have to look up selling prices of local houses, the features of your house, and the quality of the environment around the house. If you really wanted to get the most money, this process would normally take a couple of weeks.

Artificial intelligence can take all of the data found online and give you the most optimal numbers within a few minutes. Now that we have the quality of school measurements, crime statistics, online selling databases, and a lot more data, artificial intelligence can be quite the useful tool. Instead of eyeballing a number, the artificial intelligence will find the most optimal number given the data, which usually results in a higher price.

However, that's not all that artificial intelligence can do. For instance, Zenplace is an application that allows landlords and property managers to be more effective. The website offers services such as being able to find new tenants, keep optimize schedules of maintenance and repairs, and even allows tenants to pay through the chat system. An additional benefit to artificial intelligence is how much information it can memorize.

When a tenant comes to buy a house, they ask for things like square footage, the terms that the tenant will be agreeing to, and many complex things that are not easy for a human to remember.

Safety on All Sides

As for the laws surrounding real estate, most of it is just protection. There are a few different ways to utilize a piece of land, but it's usually the government's choice. Not only that, but you have the fact that different states have different real estate laws. The problem is that there is so much law concerning the buying and selling of a home that it can't fit in a single book less than a hundred pages.

A good example of this is the real estate licensing that occurs. In some states, you have to be licensed in order to sell houses. In other states, it's an optional item that adds to your credibility. A real estate license is a far more complex process than getting a regular license. You have to sign up for a pre-course and then take a licensing exam, both of which are state approved. This brings you on board to becoming a real estate agent that is licensed.

However, the benefits of being a real estate agent that's licensed sometimes negatively impacts your business. You have to disclose that you are a real estate agent, which some people prefer not to work with. You often are required to work with a broker, which represents a bigger business. This is to protect the consumer because if you don't have the money, the broker will probably be insured. This allows consumers that have been victims of fraud to go after someone that has money.

The beauty about the internet is that it has taken these laws and simplified them. For instance, one could use a website like Trulia. This website allows you to look for houses in your local area. It then provides many of the benefits that have come with the internet. You can understand the school area, crime statistics, and even get a virtual tour of the house and neighborhood. Once you've decided on a house, you can then get a pre-qualified mortgage loan to cover that house through the website. You can then contact the owner of the house and try to get the house.

The reason why I point this out is because of how much time it used to take doing all of this. It's only recently that we have graded schools. It was really only since the 1960s that we started having a robust crime statistics system. It is only within the past decade that you could get a mortgage loan and a house on the same website.

Therefore, some of the information was never available to you. You would have to travel to each house and area to do research. It might take you a week or more to do research on an area. You would then have to go to a bank and get a mortgage loan but only after you had chosen a home. You would have to look at advertisements or, further in the past, just walk around the town you wanted to be in to hopefully find a house.

Thus, taking the jump to purchase a house was usually a month to a three-month-long endeavor. Now, you can spend a few hours of your day looking at residences to find your perfect house. You can then spend maybe another 30 minutes filling out a mortgage loan application. Finally, you can then start the process of buying the house you wanted without investing anything ahead of time. All because websites and companies have simplified the laws around buying and selling real estate.

Filtering Out Time Waste both Client and Agent Side

As you can tell there's actually a common theme running through all of the evolutions of real estate. Almost all of the evolutions happened because it saved time for either the client or the agent or both. Companies formed as a way of selling faster, thus saving time for real estate agents. Advertisements came out that saved time for customers trying to find homes elsewhere in the country. The online world came into existence and it saved time for both of them by having the services online. Websites simplified the barrier for getting homes and expedited

communication between buyer and seller.

The rules around real estate haven't really evolved that

much beyond the standards. There are standards in place that

real estate agents must follow. These real estate standards have

been put in place to prevent fraud and protect customers.

However, they depend on the state you plan to sell in.

MACHINE LEARNING TO EVALUATE PROPERTY PRICES AND LOCATIONS

How A Real Estate Agent Sells Property

Crime Report Statistics

Often, the very first thing that a real estate agent looks at is the amount of crime in an area. Nearly every city has crime statistics to determine where they need to place their focus because they only have so much manpower. In the most advanced cities, these crime statistics are actually broken up into regions, counties, and even streets.

It's important to understand the level of crime that happens in a place because a tenant will actually value their safety over the value of their purchase. You might have a tenant that doesn't care about a high crime area if it's only theft and they tend to not have anything they think is worth stealing.

However, someone like a software developer with a higher-paying position than an individual that lives a minimalist life, they are likely to have a lot of money they want to spend. This means that they will often have items in their household that they don't want to be stolen. In addition to this, you might have a family who generally wants to make sure that their house is not going to be broken into. This is not only for the safety of their items but also for the safety of their family.

Quality of Schools

With families, a new item is added on to the list because families tend to look at just one more criteria. Families are concerned about the quality of the schools in the area.

Even though most public schools are considered governmental facilities, they are not operated like a government. The school in a high-class environment will often benefit from higher taxes in the surrounding area. This is because most schools are dependent on the taxes collected in that area by law.

This creates a rich vs poor school and family. Families who have a slightly higher amount of pay want the best for their children. Therefore, what tends to happen is people with higher pay go towards higher schools and rich schools remain rich schools. Vice versa for poor schools.

Median Income of Surrounding Residents

This is actually a secondary cause for something else that many advanced tenants will look for. The average cost of the house will often be determined by the average median income of people in the neighborhood. This is important because knowing that average, one can guess at the likelihood of crime and quality of the school.

If a neighborhood has a high median income, that means that they're less likely to steal or break into the other houses. Since school quality is often determined by how much taxpayer money is available, the median income of neighborhoods often determines this. This means if you really want to be looking for houses you need to start looking at medium incomes of areas. This becomes similar to an algebraic problem where if you have any two of the variables, such as median income and crime statistics, you can figure out the third.

Material Cost of the House

The last item on the list that is often requested by more advanced tenants is a request of building materials. Knowing what a house is made up of, one can begin to delve deeper into things like health and maintenance.

There are many houses in the United States of America that are from generations past. As real estate has developed, the standard at which houses are built has changed. This means that the older a house is, the more likely that this house is going to be unhealthy and need a lot of maintenance.

A new tenant will likely not want to buy a house and then think they need to repair it. It is only in situations that people purchase a fix-up house or a starter house. These are houses that people buy simply to fix and then resell. People who are buying houses that are buying their house on the third or fourth time will often ask for details on the materials in the house. This is so that they have a good understanding of just how much time that house is going to take up in repairs and maintenance.

How Machine Learning Generally Works

Input

The good thing about this is that all of these are integer forms of input. Machine learning depends on the numerical input you can shove into a system.

A machine learning algorithm doesn't work by interrogation or any other sci-fi method. Machine learning doesn't understand context, which is both a good thing and a bad thing.

It's a good thing that machine learning doesn't understand context because that would mean that we have general AI and they would be writing this book. It's a bad thing because that means if you want machine learning to work on something, you have to turn what you know into numerical data. You have to translate context into numerical data for machine learning to work.

Thresholds

Once machine learning has the data it needs, it also needs thresholds. You can think of thresholds as the machine learning form of context. The simplest form of a threshold is a binary choice, whether it is something or it is not something. There are more advanced forms of thresholds but for the purpose of this book, they are just important to understand in that thresholds are the decision makers of neural nodes.

When a machine learning algorithm goes about learning, it will take those inputs and bring them into a neural network. In the neural network, you have serial and parallel neural nodes that breakdown the inputs and build them up into a decision. Oftentimes, the inputs save the neural network the trouble of breaking down the information.

The neural network then decides whether a certain piece of information continues through the neural network. What that means is that the neural node has made a positive decision, which will ultimately affect the overall outcome that the neural network makes.

Gradient Descent

Once neural networks have made their decisions, that means it's time to see whether the network needs improving or not. Improving the network manually is not really recommended because humans are flawed and the network could become biased. Therefore, that means the network could give false positives. This is why the gradient descent was applied to neural nodes and neural networks. Gradient descent is a calculus equation that determines the most immediate optimal path for a mathematical decision.

Thus, the input is fed into the system, the network makes its decisions, and the result is called the output. Once you have output, you then use that to determine the new input or reverse that output to become the input. By having ways of telling the machine what is correct and not correct, you provide a continuous method of optimization. This forms what is known as the gradient descent.

More Data Is Useful

Obviously, the ultimate conclusion is that the more data you have the more accurate the machine is going to be. You could have a small neighborhood where you input the data and get the results you want. However, the problem lies in the fact that it's just that neighborhood.

If you're talking about data from a single neighborhood, you, a human, could do the calculations faster than it would take to build a neural network. Building a neural network for a small amount of data is pointless. This is why products have been made because products allow access to more data.

If you are a single real estate agent, you might sell a hundred homes in your lifetime, maybe even a thousand. In the most optimal case, you're looking at a thousand different neighborhoods over your lifetime. That means the data changes over time, the data is old, and the data is still relatively small in the scope of real estate marketing. By having a product, real estate agents could combine this information to produce a machine learning algorithm that is much quicker.

Machines Learning to Evaluate

A Machine Needs Only Numbers

Now, as I said, you need to translate context for a machine. However, we've already done this ourselves. In order to look at crime statistics and compare them, we have already put a numerical value on murder, thievery, and fraud. In order to look at the median income of a city, we've collected the income of everyone. To improve specific neighborhoods, we have taken the time to assign income to quadrants.

What this means is that most of the information that we needed to translate context has already been done. It's just a matter of getting access to that information and compiling the information into a machine learning network.

Numbering Context Over Years

This also hasn't been an easy process and it's taken many decades. In fact, we didn't have a really good consensus until about the 1960s when computers began to get involved. The simple truth of the matter is that humans are slow compared to machines when it comes to numerical equations. So, by the 1960s we had a population so huge that we couldn't keep track of it. Not by human hand at least, which meant that we needed to hand it over to machines.

Therefore, since the 1960s, we have quantified these contextual items into numerical form for computers. This has bequeathed us a huge library of information that real estate agents can use to make machine learning products.

Quality of Life Now Measurable

Perhaps the most advanced measurement is the quality of life measurement. It is only in the past 100 years that we have been able to adequately measure population among other things. It is only in the past 50 years we have been able to measure the quality of life. The quality of life includes things like life expectancy, income levels, poverty levels, and many other variables. This is useful for machine learning algorithms when it comes to selling houses.

Selling a house is really a combination of two different things. The first item is that you need to measure the cost of the house. Then you need to measure the cost of the environment. A million-dollar house built-in the poorest area of the neighborhood will have the poorest environment, so it is likely not going to be a million dollars.

Compensating Other Context

This creates a compensatory area where there is now competitiveness between the cost of the environment for the cost of the house. For instance, you may have a poor house, but it might be nice to live next to some of the fanciest restaurants there are. You might have a house on the water, but the house is rundown.

In other words, you might have a gray house in a very bad environment. You might have a horrible house with an environment that doesn't just include the basic necessities like crime and school quality. Being able to say that your house, that you're selling, is near a grocery store or high-end restaurant gives you an additional selling point. This can be referred to as compensatory cost or the benefits of living in the area that are not necessary.

Faster Faster Faster

Selecting Potential Houses

By having a machine learning algorithm keep track of all this, you're able to move much faster in the market. You're able to find the right houses with the right environments and buy them at the right times.

In the past decade, the average way of buying a house was looking at what was available at different times in the day. You would go over the same process and waste an hour to an hour and a half a day. This was simply to look at what was available.

A machine learning algorithm that understands what you're looking for and can do this for you. It can also do much more than that. We can find the right house for the right price but also do the three to four hours of research. Should you find a successful house, you have to do additional research.

This research included crime statistics, quality of life, and school quality. A machine learning algorithm will find the right house for you and look at all those variables automatically presenting you with the best house to buy.

Viewing Those Houses

On top of this, if you mix virtual reality into this selling process you don't actually have to meet with your clients. The more likely case is that you'll bring a client into a viewing area where they put on a headset and look at the inside of the house along with other clients looking at the same house.

This cuts down on the amount of time wasted on driving from house to house. Instead of having to negotiate times with clients to go look at houses at different times, you can simply bring them into a room where they put on a headset and look around the house as much as they want. Almost 90% of a real-estate agent's wasted time comes from driving.

Filtering Applicants

You can also begin to filter out applicants based on things like relative income. In the past, someone would apply to get a house and you would request information from them. You would then go through the process of getting that information, doing a background report on that person, and getting a credit report.

This process actually takes quite a while. In fact, while the house may be up for sale at the beginning of the month, it may not be sold until the end of the third month. This is because this communication between humans is rather slow.

You might send a request for information and wait another week to two weeks to get that information. You could then run a background check, which requires them to pay you to compensate for the cost of a background check. You then need to wait for the background check to get back to you, which has increased in speed over the years. You then also need to get a credit history, which can take some time.

Then, once you have done this for every applicant you then begin going through the applicants you want. If the list is somewhere around 10 to 20 people, you're looking at around 10 to 20 people that you have to repeat this process for. Even worse, this process is manual and so you have to personally spend time.

Knowing this, machine learning and the internet can make this process a whole lot quicker. A person can sign up to your website, submit all their information, and process a fee for a background check and a credit check.

The best part about this is that this ecosystem is all online and automatic on your side. This means you no longer have to waste time of your own to go through every manual step listed out before. Essentially, you can automatically filter out the applicants you don't want without ever having to deal with them.

Even Dotting the Lines

The best part is that it doesn't stop there because you can use online services and machine learning to formulate contracts. You can use the past history of the person to create a contract that better fits them and you rather than use a one-contact-fits-all solution. A person might have a disability or a cat or some special circumstance. You can use background information of a person to generate a contract that specifically aligns with the needs of yourself and the person.

Additionally, you can have that contract signed in a more secure way. By having the contract online, you can record the signing of the contract. This allows you to progress with the contract without the need of a notary. Setting up a notary can take time if you don't have one in your pocket. By being able to record what you're both doing you don't need a notary to confirm a contract in most situations.

A New Frontier

Skyline

Skyline is a company that helps to lower the cost and mayhem that comes with asset and property management. It primarily focuses on predicting trends in pricing and quality of property you might want to buy in a specific area. Additionally, it also schedules pricing rental values with renovation choices so that you get the optimal return on investment.

You act as a member in their program and they make money when you make money (take a percentage of your revenue). Having said that, they tend to throw around the word "AI" like it turns everything into gold. After all, you can predict the growth of population and neighborhood value if you watch a few factors. These items will increase if the following happens. More jobs often means that more people want to live there. Tourist attraction count going up or improving often means that seasonals will want to be there more. Tech innovation in the city and in commercial areas mean that business value goes up, meaning it's worth more to work in that area. Lastly, city populations are calculated per year. You can compare that with how many children were born each year to find out how many people moved to that city. This will tell you how much the population is organically growing versus economically growing.

When it is economically growing more than organically growing, that means the city has a lot of value in it.

Now, if it just seemed like I threw variable after variable up there, it's because that is ultimately what the artificial intelligence usually looks at. There may have been variables I did not account for, but most of this can be done by hand. Then you can evaluate the value of neighborhoods in the city to come to an optimal price for rentals and selling.

The important thing to note is that AI does all of this for you, which means you have more time on your hands to make money. That is the *ultimate* point of this type of software.

Proportunity

The primary goal of the company is to lend money to then sell homes they have their hands on. They do have an option to add real estate from other websites, but they need to be on popular websites that they approve of. This creates a bigger version of landlord and tenant in reality, but with much higher rates. Since this is a loan, interest rates can be charged on top so it's far more expensive. They'll talk day and night about the discounts, but what they really try to hide is their "smart platform".

AI is very similar to what the average "Gaming PC" was in 2016, very confusing for the average person. Unless you have someone explain it to you (it can still be confusing in some cases), AI is a tangled mess of foreign words. The only paddle you get to wade into the AI pond are the mathematical expressions… and those are unwieldy, to say the least.

While one cannot say exactly what their AI is doing, one can certainly speculate. There are many factors that identify growth areas and undervalued houses. The percentage of available job listings is a good way to see city growth. The percentage areas of bought homes during certain times also help with determining neighborhood growth. Undervalued houses could simply be houses that are sold by small-time sellers or independent owners that didn't do some easy math. If you take each house on a block and determine the optimal percentage those houses are sold at. Then you do this with the existing real estate in the city, you get a percentage range. This is a basic range at which you can mark up the value on a house. If you're below this range, you'd be undervaluing a house.

Now, this isn't to say that there isn't something more complicated behind the scenes. It is just very easy (as an AI programmer) to see how one could make such a system. However, if one looks at the Terms and Conditions, they make it so that no one is allowed to show anyone how their website works. That's right, they have legal terminology specifically designed to try to curtail Fair Use law. According to these guidelines, I can't even citation them. Make of that what you will.

Enodo Score

Although the website is a *little* broken, it is quite the impressive tool. Enodo is a beta website (as of writing this book) that allows you to consolidate a lot of information. The artificial intelligence seems to come in the form of a web crawler and statistical mathematics. A web crawler is a programmatic mechanism that searches the internet to discover information. This one seems to target key search terms, pre-storing the results in a database.

In the case of Enodo Score, they provide you with both the information and analysis of neighborhoods in a given area. This is actually quite expensive if you try to hire someone for this. If you do it yourself, it's also very time-consuming. Therefore, the website is really built around the concept of saving time.

Other Companies

Companies like these are really the first of their kind because of the AI implementation, but the implementation is light. This is because the bar for entry is exceedingly low right now for this market. These can be useful tools, but you could create such a system with improvements rather easily.

I see a few more things happening before this industry becomes noticeable by the average person. The average person won't use the search terms "AI" and "Real Estate" together. Here's what I see happening.

The first thing that will likely happen, since AI has already started its incorporation, is Blockchain incorporation. The Blockchain would be extremely useful for the real estate industry, especially for record keeping. Since the Blockchain is nearly impervious to tampering, it serves well in cases where a notary would play a role.

The second thing that I see happening is that real estate companies will act more like software companies. AI can handle helping people find a house, sell the house, and a human is only involved in transaction completion. Therefore, with things like viewing, it'll be done through the browser or on the phone. As such, the tools will eventually become the company front.

However, the days where a real estate agent walks you through a house are slipping away quickly.

The final thing that I see happening is something we see from the tech community: material reuse. There are a lot of homes being constructed, even though the birth rate has actually slowed down and the death rate sped up. This means that real estate building will inevitably decline. Even more, many are now opting for the RV life, a real estate opportunity of its own.

It is not only cheaper to update an older house to newer standards, but it's less taxing on the environment. It's less expensive and it's easier to do, thus recycling houses will become even more common. As there are fewer new families, there will be a natural decline in new houses once the market figures this out. As of right now, the market has yet to figure this out on a wide scale so only small companies are doing this. However, those companies are quickly growing while "new house" companies are flopping quicker.

With applications like Airbnb taking advantage of this overall trend, they've quickly become one of the biggest markets in real estate rentals. There will be similar companies climbing that ladder using different methods; AI methods. Thus, it is only natural that once Blockchain is incorporated and real estate becomes more like software that real estate will become a reused resource like t-shirts.

HOW WILL REAL ESTATE AGENTS BE AFFECTED

One concern with the inception of Artificial Intelligence is that we will replace all real estate agents and contractors with robots. While this is very unlikely to happen in the future, there will definitely be a change in the role that real estate agents play in the process; and those who fail to adapt to new processes might be rendered irrelevant. We investigate this issue more in this chapter; looking into the role AI will play in the process, and also check out a few exciting new companies in the space.

Real Estate Agents Will Never Fully Go Away

People Always Get People

When people talk about artificial intelligence taking over a specific position, they tend to forget that people like talking to people. The average real estate agent's job may change, but there's always going to be a need for people.

The best example of this is actually McDonald's. McDonald's has quite a few locations where you can use a screen to order the food yourself. Yet, even though this option is available, they're usually quite a few people at the cashier counter to take people's orders. This is because people like communicating with people.

The internet itself is a recognition of this because it's one giant machine used for communication. Literally, the internet is nothing but a communication tool. It's just that through communication, we can order items internationally, we can talk to people in different languages in other parts of the world, and we can collaborate on projects better.

Machines Cannot Understand Context

Real estate agents will always be needed because sometimes a computer doesn't understand the full situation. As I mentioned before, an artificial intelligence that is capable of understanding context will replace every single person's job. We simply don't have that, we have a clever mix of mathematics with people who translate context into numerical values.

This means that if a person recently came into some money, but they were previously poor, it would skew the data. You could have a rich millionaire that was poor for 20 years of their life that wants to buy a house from you. If you solely relied on machine learning, this would be an opportunity that you have missed.

The example of what can go wrong might be to the extreme, but there are many situations like this where human eyes are needed. Machine learning doesn't replace work, it minimizes work. It gets rid of the manual labor that most of us experience. Therefore, while you don't necessarily want to fully rely on machine learning, machine learning is still very beneficial.

People Play People Better

In addition, people upsell to other people more efficiently than machines. Machines have gotten incredibly good at selling certain items to people, like loot boxes. However, a salesman will always be able to outsell a machine because the salesman gets the context. There are certain situations where machines can do better, but overall, people can't read other people.

Let's say that you have a young person escorting their disabled grandparent to buy a house. The machine learning algorithm would try to sell the house to them based off of both of their wages, but a human might try to sell a house based off of the young person's worth. You might be thinking, why not go after the older person? The older person is likely to have more money, why not go after them?

A real estate agent will look at the finances of both of them and come to a conclusion. The healthy young person is more equipped at continuing the sale past death. Instead of trying to sell the house to the grandparents, a real estate agent might suggest putting the house underneath the younger person's name to ensure that the house isn't put on the market should something horrible happen.

This would be a situation where the younger person makes more than the older person but has less financial responsibility. Therefore, the real estate agent might want to try to lease to the younger person in a situation like a rent-to-own. Instead of selling the house, the context of the situation calls for a much longer-term relationship to ensure that the real estate agent gets the most money out of the deal.

If the younger person fails to pay rent, they don't have to claim bankruptcy. If the younger person owns the house, to get rid of the house they have to claim bankruptcy. In this situation, it is more ideal for the younger person to go with a renter situation. For the real estate agent, if the younger person takes the deal, then not only do they still get money towards what they would have made of the house, but they get something more, potentially. If a person fails to pay rent, the real estate agent made tons of money while they were renting and now they can resell the house. So, given the contextual situation, the real estate agent has an opportunity to make more money in the long term.

AI Became the New Frontline

AI Filters Applicants

In the past, applicants would send an email or postal mail to try and buy a house. In the postal mail days, this would often take weeks at a time. It would take weeks because even if you sent out the mail in your city, you likely wouldn't see it in the mailbox it was going to until about 2 to 3 days later. Trying to send that out to another city, you could expect about a week or two before it got there. Emails made this whole process significantly faster.

However, both had the same problem attached to them. Let's say that you got 50 applicants in a single week. It takes you about 10 minutes to look at the application that sent it. From opening it, to reading it, and to deciding on it, all of that took 10 minutes. That means that mail wasted 500 minutes of your time. That means that nearly 9 hours of your time was wasted on just mail.

On the other hand, email wasn't as efficient as everyone perceived it to be. Yes, you are likely to send and receive the application faster but going through it is the same speed. It also gave you an additional problem; more people. By receiving the mail faster, you received more applications faster. Therefore, instead of 50 applicants in a single week, you might have gotten somewhere near a hundred.

So, you have more customers but you're wasting almost double the amount of time. You're wasting 17 hours on just reading mail. Now, that may not have been the experience for all real estate agents, with some being worse and some being better, but it's an example. This highlights a huge problem.

The problem is that there's simply not enough time to look at everyone's application in a single week. This is why even though communicating that application has been faster, going through that application as not really improved until recently. This is because artificial intelligence can be used to sort through those applications.

Artificial intelligence allows you to select parameters for what you look at in your applications. For instance, if you want people of a certain level of income for specific houses, artificial intelligence can sort through the email. You simply request the annual income when they try to buy a house. The artificial intelligence looks through all of the emails and associates that income to the house to see if it meets the requirements. If not, that applicant is automatically filtered out or is put in a separate folder for later consulting.

As you can see, this allows you to automatically sort through the applicants you would have naturally rejected.

Therefore, you likely saw 100 emails a day that got shortened down into maybe 20. This saves a significant amount of time especially since it's automatically sorted for you for the best candidates.

AI Filters Agents

Knowing that you're getting the best candidates for your houses doesn't mean you're going to get the best sales. After all, artificial intelligence hasn't quite made it to a level where it can sell the house. Sure, you can definitely sell the house with artificial intelligence, but you won't get the most money out of it. The reason behind this is because artificial intelligence chooses the optimal value for the house. Real estate agents go for the highest number. That means that while the house is based in some reality, the price doesn't necessarily match it.

However, if your artificial intelligence creates a personality report of your consumer, you can have the most optimal real estate agent. When you have an artificial intelligence like a chatbot converse with a customer, they give away a little bit of their personality. Normally, when you ask for life goals or the job that they do, they'll accompany it with descriptions. Those descriptions can create assumptions about the type of character that person is.

There are cases where you do have the ability to have them take a personality test. However, that's a little bit more intrusive than some companies would like. Therefore, if someone's life goal is to enjoy time with family and they are particularly old, they are very likely going to pay attention to details. If the person is wanting to move down there just for tourist-like reasons, they're not likely going to pay attention to the key details. The key details are what ultimately decide how high the price can go. Therefore, if an older person is looking in the market, they are likely to have had several different houses with benefits. If the person is a tourist that is looking for a permanent residence, they're likely looking for the easiest and most outwardly beneficial place. This means that the real estate agent can sell higher to the tourist.

There are many situations where you would want to create personality profiles on the fly, but you wouldn't have a lot of information. This is why chatbots are useful because a chatbot is a lot less intrusive than a personality quiz. You can also word the questions much like a real estate agent might ask a question, which hides the true purpose of the chat box system. Using artificial intelligence, none of your company's time is taken up by assessing the personality that this person has.

AI is Faster and Can Do More

Lastly, artificial intelligence is extremely good at doing things really fast. It can look through hundreds of records of history to find the best houses. It can look at the income levels of an area versus the crime statistics to choose the best environments. It can throw you forecasts of how much the house might be worth in about a year and do so much more. The point about this is that it is much faster for artificial intelligence to do.

Now, you might be thinking about how you can easily just go online and look up the level of income for the county. This is quite easy and many real estate agents do it. You might also be realizing that most of the steps are repetitive and easy. If something is repetitive and it comes in numbers, it can be done by artificial intelligence.

Let's say that your company has 10 houses a week to go through. Each house must have a profile attached to it, which means school quality, environmental quality, and structure quality. Environment quality is broken up into crime statistics, job availability, and convenience availability. Structure quality is broken up into assessments, appraisals, and improvements. These all take time and almost always follow the same steps. Artificial intelligence can do all of this. If you're spending somewhere around a work week just to maintain this, you can put an artificial intelligence to task. This is where you can minimize the amount of work done by humans, thus saving you money.

Where to Minimize Work

On Application

The first way that you can minimize work with artificial intelligence is in the application process. The application process is the slowest of all the processes in terms of bulk work. You might have to wait a bit longer with background checks and credit reports, but the application takes the longest physically.

The first bonus is that we can have these applications online. That means somebody goes to a website and fills in their information to apply. This saves a ton of time because it used to be that you mailed your application in, sometimes still do.

The second bonus is that these applications are often stored as email. This has worked well for a long time but sifting through those applications takes a long time. Therefore, while we don't have to wait for postal mail, we still waste a lot of time during the review process. This is where artificial intelligence can minimize the work. Artificial intelligence can take the application, review it, and sort it. This means customers hear back almost instantly if they are pre-qualified.

On top of that, artificial intelligence can take what's inside of the application and use it to find available houses. This would mean that the artificial intelligence provides you with houses alongside current applications. On top of this, artificial intelligence can submit background checks and credit reports. Since most of the information on an application is the same information on background checks and credit reports, that information can be reused.

This means that a customer can apply, pay a fee, and you receive everything you need without having to manually review anything. You get the background checks, credit report, the application, and the houses that this person can participate in a transaction with. This saves a real estate agent a lot of time.

However, there are other aspects of the transaction that time can be saved on.

Initial Interview

This is where the initial interview comes in because once all the information is in, it's time to meet the client. The problem is is that this is more of an introductory interview than anything else. Most real estate agents have a transcript of questions they want to ask their prospective clients. This means that they are going to ask the same questions to all the clients with only some of them having additional questions because of certain elements.

An artificially intelligent chatbot system can handle this interview for you. In fact, right after they put in their application, they can continue right into a chat box. This chat box will ask the questions you plan on asking them. Likewise, the prospective customer can ask the chat box commonly asked questions. This saves you the time from having the initial interview and the time used for them to ask questions.

86

In addition to saving time during the initial interview, this area can actually be optimized to include personality reports. Personality reports provide you with a way of figuring out how to sell to the person. Half of the battle of selling a product is selling yourself to the person you're selling the product too. Therefore, most of the sale pitch is built upon how you can figure out their personality to figure out what they want. This is because that is how most sales are done, regardless of what numbers are in front of a person.

A person could very well who wants to buy a less expensive house, but you could convince them otherwise. You could tell them that the less expensive house has problems with it. You could also tell them that the less expensive house is not in a safe neighborhood if they have children. These are personality traits that can be manipulated to ensure a higher price tag comes with each transaction most of the time.

Setting Up Meetings with Potentials

So, you've selected your candidates, they've chosen the houses, and now it's time to set up meetings. Well, artificial intelligence can handle this too. It used to be that if you had too many clients you would hire a secretary to handle meetings. Now, you can have artificial intelligence set up the meetings for you.

You simply give the artificial intelligence the standard routine that you follow. You add any item that you need to do in addition to that meeting. Then the artificial intelligence takes all of the timetables and provides the client with available times to have the meeting. There's no need to interact with the client and everything can be done with a chatbot.

Not only that, but you can also have artificial intelligence reschedule meetings. If you have to insert something or another meeting took too long, it is a button press to reschedule. The chatbot contacts the client and renegotiates a new meeting time with them.

Paperwork and Auto Filing

Let's say that you finally finish getting your customer to by the house, what about the paperwork? Well, artificial intelligence can help with this too. Most of the agreements that come with the real estate agencies are repetitive in practice. You have a party that is buying, a party that is selling, and a party acting as an intermediary. This means that you can turn artificial intelligence on to this.

Otherwise known as smart contracts, not to be confused with contracts made with blockchain, these contracts are developed by artificial intelligence. Therefore, you simply enter the information of the client by simply clicking on their profile. You click on the seller of the house. You then auto-generate a contract based off of the house's details. You, of course, review the contract to make sure that it is a contract that fits your needs.

This is very important because contracts can either be cookie cutter formatted from lawyers or specially tailored. Needless to say, these contracts usually require the cost of a lawyer. Hiring a lawyer for every single contract that you need is not ideal. By having artificial intelligence auto-generate the contract, you save time and money. You can then just send the contract off to whoever needs to sign it. You can even do this from your phone in most cases.

Why This Environment is Worth Investing In

It's Still a Small Environment

The benefit about all of what we're talking about here is that it's still a relatively small market. There aren't a lot of people in this market because it doesn't receive a lot of notoriety. The real estate market has had a very slow overtake when it comes to technology.

This means that there's plenty of opportunities to grow because real estate agents often love these tools. The best part is that you can actually help out another industry in the process. You see, there are tons of Junior programmers left jobless because most programming positions need a higher level. However, many of these applications are something that a low-level programmer can handle.

The artificial program needed to run the chat box and filtering system can be handled by early programmers. This means that you can hire them at lower rates than otherwise expected and still receive a mostly adequate product. So not only can you use these tools yourself but, if you make them, you can also sell these tools rather quickly.

It's Useful and Saves Time

The reason why this is an investment opportunity in both the consumer and the producer side is that it's useful and saves time. It requires no long-term liability to start because all you need to do is hire a programmer. A programmer is responsible for the artificial intelligence, the websites, the applications, and even the product design if you have one. This opens up the opportunity for not only a hurting industry but also a new industry.

I imagine the new industry would be underneath IT Room and Board Maintenance or something similar in nature. These would be positions where maintenance people are combined with IT professionals. Therefore, since once the application is made and all you need to do is maintenance it, it becomes a two-in-one job. In fact, most real estate agents hired somebody to do their website and now they can rehire that person to make this. That doesn't necessarily mean that you need to hire someone to do this because there are products, but this is an opportunity to sell a product with a low entry bar.

It's perfect right now because it's useful, it saves time, and it's relatively new. Being relatively new means that there's not much you have to do to become a sellable product. For instance, simply providing a platform where emails are filtered by artificial intelligence and then sorted is a sellable product. Creating an application that allows you to expand security (facial recognition cameras) is a sellable product. It's also not as difficult as it sounds.

There is a ton of documentation on facial recognition and plenty of companies that will rent the service out. You simply combine that with a monitoring program that looks at all the different cameras and you have a product. It's a really easy product to make yet no one is making it because there's not a lot of attention. You provide jobs for programmers, maintenance people, and are able to sell real estate properties faster.

Long and Sustainable Growth

The beauty about this is that once the application is up-and-running, it usually doesn't require a lot of costs to keep running. If you think about it, you might have to hire a developer for a couple of months. You are likely looking at a cost of about $10,000. This is a standard investment when making any sort of web application so it's really not that much. Once the application is made, not only can you use it, but you can also sell it. You can have user accounts, limited subscriptions, and offer it as a service.

The cost of upkeeping such a website is relatively small. It's relatively small depending on how it's built. However, you are not exchanging that much data, so it wouldn't chew through bandwidth like videos do. Artificial intelligence would work on the server, so the server might need to be a little beefy, but not at first. Especially if you are renting artificial intelligence from someone like Google. So, what you have is a monthly bill for the server, an annual bill for the web address, and an investment in a programmer for a couple of months.

After that, you only have a bill for the server and an annual bill for the website. The cost of running the website would probably run you between $50 to $100. Yet, if you sell it as a service then you're likely going to make maybe $20 per client. Perhaps more depending on how effective your software is. Your upkeep is not only paid for by a handful of customers but you're making a profit off of a relatively small product. This means that over time you can add additions that you want and then pass the cost of making those additions to your customer. Therefore, you get a better product for yourself that you then get reimbursed for by the customer.

Because the cost is so low, nearly anyone can do it if they're in the real estate business already. This is because $10,000 is practically nothing when it comes to investing in real estate.

Rex

Rex is an AI that primarily deals with self-assembly. REx is as close to General AI as we've reached so far. General AI is a form of AI that gets its name from the original computers in the past. In the beginning, computers were tasked at doing one thing and one thing only. Perhaps the best machine to represent this is the computer that broke Hitler's Enigma, the machine responsible for encrypting messages. This machine couldn't play video games, it couldn't even do basic mathematics. It was a one-purpose machine, the rotate through letters faster than a human to decrypt messages.

It took a couple of decades and smaller parts for humans to put tasks together. This is when the computers we often know as those big expensive ones that used to be used as businesses. You would make a program by punching holes into sheets of paper. Eventually, it was consolidated into using typewriters to translated natural language (like a for loop) into something the machine could understand. As more parts came out and better coding schemes were adopted, we started to see computers the size of today's servers, then desktop computers, and, finally, laptops. Now that we've been doing this for ½ a decade, we have computers in our pockets that could trounce that first machine.

AI has evolved in a very similar way. The first AI was built on Slope-Intercept Form and from there, people started creating scenarios with yes and no themes. They eventually found that if you combine two of them, they could handle different scenarios. This would be like going from the single-purpose computer to the combined task computer. This has evolved to break down language to give us voice typing, AI in video games, and all the things we know as AI.

However, just because a computer can do multiple things, that does not mean that it can think on its own. There's a ton of information I am glossing over when I say this, but everything you do on the computer has been pre-programmed before User Input. We use clever tricks to make a person think the computer is doing more than one thing at once. Just as in AI, we had a single neural node so too in computers did everything run a single core. Additionally, with a neural network, we created multi-core systems that *truly* do more than one thing at once. However, the OS is still, primarily, a single-core process. It cannot think on its own, it has to be pre-programmed.

General AI, like REx, is capable of defining its' own process and thinks on its own. Yes, REx is the first General AI. Not because it can make choices for itself (like many famous AI robots) but because it can do so without human input.

REX, the real estate company, on the other hand, is very similar to companies we've talked about before. I could explain what they do, but I'd be mostly repeating myself. However, they do do something different than the others. Specifically, when they match sellers and buyers, they do so by matching them on other websites. Therefore, if a seller is on Zillow and a buyer is on Facebook, the machine learning algorithm will connect the two together. This opens up online selling to include all of the popular online real estate websites rather than confining sellers and buyers to one. In addition to this, the REX agents are salaried stuff, which generally means much better customer service.

HOW REAL ESTATE OWNERS CAN USE AI

An Entry Barrier

Remove High Debtors

Debt can be calculated by a few different things. It can be a past history of debt, a current timeline of net revenue, or it can be future debt like a student loan. We all know it is important to know about debt so as to prevent ourselves from getting into bad contracts.

The problem is that people generally don't like giving out this information. They want to keep tricking people into contracts as it is financially beneficial for them. People like this have common elements to them though, such as having multiple credit cards and subpar credit history.

Artificial intelligence can look at a credit report for you and simply spit out a score for the person. Therefore, if the person has a lot of accounts open but all of them are empty, they probably deserve a good score. However, if they are empty and there's a bankruptcy on the credit report, then they probably deserve a low score. Looking through and understanding a credit report beyond the score can be done by a human. However, artificial intelligence can do it faster and this is important. It's important because you will likely have far more than just one person's credit report to decipher.

This really depends on how invested you are in credit reports and background checks. Some people like to look at a credit history with a glance and some people really dive in. Artificial intelligence would be able to save time for the people who really like to dive in.

Remove Risky Behavior

Now, it's very easy to remove people based on their smoking habits, their choices of cars, and similar decisions they make. In fact, it's kind of binary in that way.

However, the real estate industry could benefit from a tool companies use to hire people. Giving people a personality test off of your own biases, you can assess whether a person is going to be a risky investment according to your criteria.

Everyone has their own unique way of deciding their investments, but usually even if you're looking at a company you look at the character. The character of a company will often tell you the likelihood of success, especially in failures.

Real estate agents get a lot of sway about what they can do when it comes to judging who they want to sell a house too. This benefits them when it comes to personality tests, which can be taken online.

A personality test not only judges the character of a person but also tricks people. You see, companies know that people will sometimes lie in order to get that job. Therefore, as a defensive mechanism, companies use lures to bait people into choosing what they want them to choose. For instance, if you have an employee that secretly takes brakes to smoke, what would you do? In this situation, a company will give you about five different ways you can answer. However, the important bit is that you pay attention to the two answers that sound most like each other. One will sound like what the applicant thinks the company wants to hear. The other will be what the company wants the applicant to do.

You can have red herrings in a personality test and they help determine whether that person is going to lie to you or not. Knowing the percentage at which a person is likely to lie to you, you can then remove that risky behavior from a contract.

This percentage is calculated by not only using the ratio of lure answers to the correct answers to the wrong answers but by other answers. There will be more than one person who takes this personality test. Comparing the rate which lure answers are answered, you can determine which candidates are more likely to lie to you. You want to remove the common answers that act like lures because that means the applicant doesn't see it as the "obvious" answer most of the time. Instead, you want to pay attention to the ones who mostly answer using lures because those are the ones not being truly honest.

You can do this by hand, but you run into a situation where you can only do it for a certain number of applicants. To have this tool used constantly, you need machine learning. Machine learning will refine the answers so that you don't have any common answers that work as lures. The purpose of machine learning is to optimize what you're trying to go for. Therefore, machine learning can take these personality tests and optimize them to find liars quicker.

Sadly, the goal of a personality test is to find the most virtuous person on paper and not hire them. This is because once you reach the end goal as the complete virtuous person, you have fulfilled a contradictory and impossible character the company created. In other words, it is the company's way of filtering liars from genuine people as well as rule breakers (the common conception as to why they are used).

Auto-process Criminal Backgrounds

Doing criminal background checks is a rather repetitive process. Furthermore, checking a criminal background once you get the report is a repetitive process. Nearly all real estate agents have a list of crimes that mean they won't do business with that person. Nearly all real estate agents have a forgiveness time for certain crimes. For instance, burglary can be forgiven after maybe a decade if they have a pretty good financial history.

The best part about machine learning is that it can remember your actions to define rules. You can go through your criminal background checks like normal with the machine keeping track of the differences. Once you have a sample size of about 50 to 100 criminal background checks, you can then begin testing the machine to see if it learns how you judge.

The machine will take all the differences between all the applications and figure out a likelihood of what you will choose. This means it will pay attention to your forgiveness time as well as your unforgivable crimes.

By doing this, you can create an automatic process. This process will take in the necessary information for a background check. It will then submit that information to get a background check. It will then make its own estimation on which profiles are more likely to be chosen than other profiles. Therefore, instead of sifting through hundreds of applications at the end of the week, you're likely only going to be looking at 20. Once you have the process up and running, such a system, given the example, will save you from looking at 320 or more applications a month. This saves hundreds of billable hours you could use elsewhere.

Split Between Savory and Unsavory

The best part is that you don't have to get rid of the people you are most likely going to deny. Instead of trying to find the most qualified candidates, you can have the most qualified candidates separated from the least qualified. Therefore, if you're most qualified candidates are not something you can advertise to, you can turn to your lesser qualified.

For instance, if a customer is looking for a house that cost a quarter of a million dollars, but you don't have one, you could attend a different customer. That different customer might only be looking for a house $120k as it is more common. This allows you to have a constant flow of money coming in.

You can also use the personality test to filter out people in your lesser qualified area. This would be so that if you do have to turn to the lesser qualified applicants, you are going to get reliable ones. Rather, more reliable ones. Essentially, by having machine learning do the bulk of the work for you, you're able to get the max number of customers possible and mostly focus on just selling properties rather than doing paperwork.

More AI Higher Value

Nest AC Benefits

Perhaps one of the best money-saving items you can get is an AC controller called Nest. Weather and the AC controller that predicts what you want when you want. For instance, in my house, it normally stays a nice 76 degrees F until someone gets home and then it turns to 64 degrees F.

This represents a repeatable pattern but there come some problems with it. Normally, the process happens without a hitch. However, there are some days where the air is completely forgotten. This means that instead of paying for a house that has ambient temperature as to the outside, I am now forced to pay for something 12 degrees cooler.

The reason why I mentioned Nest is that it does something that many of the other ones do not. Nest learns the habits of the residents. I do believe that there is a scheduler, but Nest will learn the times at which you commonly change temperatures; and implement them in an automated fashion.

There are some versions of Nest that will modify how much air is dispersed given the house layout. There are some other versions of the Nest that will only cool specific areas in the house. For instance, when I turn the temperature down in my house it is often because another person who prefers lower temperatures is coming to my house. They have a specific room in which they spend most the time, which means that, if I wanted to, I could have it cool only that room.

This is a huge selling point for many future customers and it is only about $200 of investment. The unit itself usually goes for about $200 at the time of writing this book. Yet, it can potentially save your tenant hundreds of dollars over years. This is because it does gradual cooling instead of forced cooling. In a normal AC unit, when you change the temperature you are forcing them into a hotter environment. This requires a lot of energy.

Nest measures the average temperature and gets a weather forecast for the house. Based on that information, it chooses the optimal time to start cooling down. This saves in enormous amounts of electricity depending on how unstable the weather is.

This becomes an advertising point.

"Automatic AC control to help you save money on your electrical bill! Come rent with [company] today!"

Not only this but if your customer's electrical bill is part of the rent because you own a complex, it becomes a different story. In such a story, you can charge a default rate for electrical energy based on the AC units you buy. As many people tend to conform to a little bit colder than what it is outside on average, you can make extra money. You do this by asking for their working hours. You then ask what they generally like the temperature as. You then set the temperature for them based on their working hours so that they have the optimal temperature at home while you save money during the day.

AI Smart Cameras

We have facial recognition on our phone nowadays. What few seem to realize is that there are locking mechanisms that double as a security camera out on the market. In such a system, you scan in a face and that becomes their key. This allows you to have more security on the keys to your tenant's rooms, but also more security between rooms. Each camera could be assigned to recognize the tenant and yourself (maybe some staff) so that everyone who should have access as access.

Now, when someone else wants to get into an apartment, they must get a temporary key. This temporary key can have a digital deadline on it to make sure no one gets a key they're not supposed to have. Burglars will now be caught in the act on security camera from the get-go and if they show their faces, they can be recognized by the AI. If it is a local tenant, it can be logged into the system. If it is a maintenance person that stole something while they were working, their access to that place will be stored via the camera. If it isn't anyone local, then you have cameras that take a picture of the fleeing vehicle as well as a picture for the cops to use.

The best part is that such an investment is an investment that improves over time. Not only that, but you can place it as a security fee as part of the rent. Odds are, such as facial recognition technology is going to have a monthly cost. You can pass that cost off onto the tenants as either a separate charge or keep it as part of the rent and just say that security is part of the rent. This allows you to make the investment, save money on the investment, and improve the security of your complex while creating an advertisement point.

This isn't only technology for apartment complexes but also for houses. Having facial recognition for a house means that a little kid won't get locked out of their house and they don't have to worry about losing a key. Having a camera that works as a key and as a security camera has many benefits in many applications for real estate.

Tenants Enticed by Live-in Tech

If you say that you have an Internet of things residence, it is a natural enticement.

People love things that make their lives easier but cannot often afford them. If you combine the current prices as of writing this book of Alexa, Nest, internet cameras, smart plugs, smart lights, and smart appliances you're likely looking at an investment of around $10,000. However, if you're selling a house this can be a bonus edition for $30,000. However, the true incentive comes with renting real estate.

With renting a real estate, these investments justify normally unreasonable rent. Allowing a consumer to set up with these Internet of Things appliances will save them money. You can then take that money as additional rent. So, smart lights, smart plugs, Nest, and smart appliances save on electricity. They save a lot on electricity, which means the money from that electricity can go towards additional rent. If renting an apartment in your place cost $800 a month plus utilities, you can then justify

$1,000 a month plus utilities. Essentially, so long as you tack on between $300 to $500 to the current rent, you will have paid off this investment in 3 years and have it permanently with improvements.

As mentioned previously, if you have a smart security system you can now charge for that as well. Additionally, a smart security system will always receive improvements to be better at security.

AI Keeps Track of Keys

If you don't want to go the route of using facial recognition as a key, you can have smart keys. Instead of having a key on a ring, this key can be easily replaceable and be even more secure. RFID chips have improved over the years and smart keys can now allow people to access more than one room.

Therefore, if you really want to get some money out of your real estate, you can begin monetizing your real estate amenities. If your real estate features a gym, the smart key will

114

allow them access for a fee. If there's a pool, there's a fee. You can even get wild with it by setting up subscriptions so that people don't pay individual fees and have a certain amount of times they can do things.

This can be used to create incentives from the maintenance people working on your place. If you have a pool or a game room, you can say that the person who cleans the most amount of rooms will get one free token to enter. This creates a reward system where your maintenance people work as hard as possible to get the reward. This also allows you to filter out maintenance people who want to do the bare minimum.

There's a lot of stuff that you can do with a smart house and an Internet of Things ecosystem. People are enticed by these systems because these are items that normal people don't buy every day. It also entices them to stay longer because maybe you don't have the cheapest rent in town, but you do save them on electricity because of these appliances. So, you may charge $200 more than the average rent but your tenant pays $60 in electricity versus the previous $150. It creates a system where leaving is tough to financially justify.

Keep Becomes Easier

The Roombas of the World

It doesn't happen very often but there are some rather nasty tenants out there. There was a couple that rented an apartment once and by the time they left, the entire apartment was covered in cockroaches. This was because they would never bother to sweep or vacuum the apartment. They had all the counters clean, all the furniture clean, the walls were fine, but they had no tools for cleaning the floor. They never bothered to get those tools either so when they left, it was a nasty surprise.

Relying on tenants to keep the apartment clean is not a habit that some landlords like to keep. However, with the cost associated with hiring a cleaning team, many cannot afford to do it and have reasonable rents. However, artificial intelligence can

116

step in once again with the fantastic Roombas. A Roomba is a robotic vacuum cleaner.

A Roomba can vacuum on both carpet and tile flooring. It uses several different types of sensors to ensure that it doesn't get stuck and that it vacuums everything. The investment is about $200 to $300. Compare this investment to the cockroach filled house and getting it back to living conditions was nearly $1,000. This very simple investment can save you tons of money with rentals and it can be an advertising feature.

Trash Recognition

Another feature that comes with the Roomba is the number of times it encounters some dirt. It keeps track of the number of times that it must clean up something, which means it is a measurable amount. This measurable amount can inform you as to whether you need to keep an eye on certain tenants within your real estate.

Therefore, if someone is being dirty you're likely to see dirt events in the hundreds on a weekly basis. This means that they are not cleaning up after themselves and they're dropping a lot of crumbs onto the ground. In addition to this, it also records when it gets stuck on items. Therefore, if they tend to leave their clothing around or other items, it will get stuck and report that. These are all measurable items to help you monitor those that do not do a good job of staying clean. This is all to avoid your house becoming a bug-infested haven.

Flexible Maintenance Monitoring

Instead of hiring a cleaning team every day to keep basic maintenance, these devices allow you to allocate that time to maybe a weekend. Perhaps every Monday a cleaning team comes in to do a better job than the Roomba can. The most likely case is that you don't even need to bother with a cleaning team but rather a single person going around. This person would be responsible for keeping maintenance on the Roombas and inspecting the homes or apartments.

What this does is it allows you the never before obtained ability of monitoring maintenance flexibly. By knowing which houses are problem houses, you can begin the process of evicting those tenants that do not obtain better habits. In addition to this, problems with inside homes become prevalent a lot quicker. This is because you have someone going into homes whenever the Roombas have difficulty and it could be clothing or it could be a leak.

Spotting Leaks Faster

Besides bugs, the other biggest problem that happens in rentals is a leak. For instance, I remember a specific townhouse that had a water heater bust. The wall along the back side of the water heater had a hollow bottom. This meant that when the water heater busted, the water went from the water heater room to the bathroom on the other side and into the laundry room. In addition to this, the tiles in the living room area raised a good foot above the rest of the tiles.

Luckily, the tenants caught it quickly and a massive amount of damage was prevented. However, if the tenants weren't home at the time that the leak happened, the damage that could have happened is incalculable. There is a way to detect when leaks are occurring and it's through the water pipes itself. Water pipes have a consistent level of pressure in them to not break. Therefore, the easiest way to detect a leak is to detect if there's less pressure than normal.

However, there are several places that need extra monitoring. Any sink needs an extra sensor because the water goes in and out from those locations. Therefore, one needs a moisture sensor underneath the sinks and a pressure sensor on the water heater. These can then be wirelessly connected to a monitoring application that is powered by artificial intelligence. Anytime the sensors are outside of the normal range, the artificial intelligence can notify you and the tenants. The tenants can then inspect the problem if they are available. You can prevent potentially horrible amounts of damage to your real estate.

Damage Reduction

Security Lessens Break-ins and Property Damage

One of the obvious points when it comes to reducing the amount of money a property causes is burglaries. While burglaries are not necessarily something the average person thinks about, you should think about it when dealing with real estate properties.

Burglaries tend to happen because of one solitary rule; the house is easy to break into.

If you have a simple door lock preventing people from getting into your house, it's like not having any security at all. You can go to an online store and can get a dummy key that, so long as they follow some steps, a burglar can easily get into a house. However, if you have a smart key that has a chip in it, that technique no longer works. if you have security cameras with facial recognition as your teeth, you don't even need a door handle in the first place. Although, it might be ideal in emergency situations. Situations such as the power going out or the camera not working correctly. Security is what prevents the average burglary from taking place in the first time.

Security doesn't need to be that advanced, it just needs to be obvious. In fact, most burglaries are stopped by simply having a fake camera in plain view that looks like a real camera. However, there are ways of detecting whether a camera is on or off. This is primarily due to user negligence because we don't want to bother checking the TV if we can see a light on the camera.

Now, there have been improvements to cameras, but one massive problem is wireless cameras. The most advanced cameras are either still working with wired connections or wireless connections. Wireless devices always send out a signal and this means that if you have wireless devices, they can be detected. Depending on how often you update your firmware, they can also be hacked.

There are plenty of technical problems with having cameras because they are made of technology and improvements happened on both sides. Every time you make a better camera, a criminal eventually figures out how to trick that better camera. However, there's a key point here that you need to understand.

The criminal that can trick that camera is often not going to bother ransacking the place. This is because if they can easily get into a place, they will try to go for that place. The easier a place is to get into, the more enticing it is. Therefore, if you display cameras, almost 50% of the people who would have ransacked the place avoid it. If you put a digital key lock on the door or a Smart Lock, you get rid of another 25%. If you put reinforced windows, you get rid of another percentage. Every improvement you make to a location in terms of security translates to fewer people looking at ransacking the place.

Therefore, there is an incentive to increase the security to its maximum but not go crazy with it. A standard house does not need a steel door coming down in an emergency. They simply need to have a system in place that makes it very difficult to both get in and escape. By having such a system in place, you prevent break-ins and thus you prevent property damage.

Security Helps Catch Responsible Parties

To further boost the necessity of security, one of the other items that Smart Security helps you with is proving the identity of the responsible parties. Even thieves have smartphones, even thieves accidentally connect to a wireless network. By having a security system in place that's smart, even if a person breaks in the alarm can go off. This is because with security systems built to recognize people also recognize phones (and other items) so you can begin to recognize strangers in the house. In a smart home, when a stranger enters the house through a window they're likely not friendly. If they are pick locking the door, they are likely not friendly. These are easy situations that machines can look at and turn the alarm on.

However, such security systems can do something that past security systems couldn't do. Everything I've just mentioned, from strangers entering the home, strangers going through the window, and strangers picking the locks, can be detected by basic devices. Motion sensors, tamper-proof locks, and wheel window sealants are real-world items. What a Smart House does is send the owner a message that something strange is taking place. The owner then opens an application to view the house.

If there is someone in the house that's not supposed to be in the house, then the owner can press the button to dial 911. This means that no alarm will go off to warn the criminal that any alarm is going off and the police are on the way. Part of the problem with old-style security is that it warns the criminal that they have a limited amount of time. This means that the criminal can plan their escape and if they know the average response time of police officers, they'll usually get off scot-free. If they don't know an alarm is going off, the police will show up while they are still trying to steal stuff. This significantly improves the ability to catch responsible parties when it comes to stealing.

Security Proves Liable Tenants

Security proves a lot of things with criminals, but it also proves evidence in cases against tenants. Very often, tenants try to say that damage was already there or that damage was not the responsibility of themselves. For instance, if a tenant accidentally falls against the wall and puts a hole in it but then covers it up with putty, only the tenant knows what truly happened. Since it looks like a half-attempted repair, no company can back up the real estate owner when they accuse the tenant of causing damage.

By having security cameras in the home under the guise of keeping people in the home secure, you then can prove that someone was responsible. Therefore, when it comes to arguments about who did what damage, it becomes obvious that such a tenant can't fight against it because you have evidence. Not only do you have evidence, but you have video proof of what they did. This would prevent a lot of situations where the damage was in the thousands.

Let's say that some kids decided to be kids and broken water pipe in the house, causing a lot of water damage. In a house that doesn't have a security system, this causes a problem because the bike could have busted on itself. There are ways to prove that the pipe was broken due to something other than the pipe, but there's no way to prove the children did it or that the family was responsible. By having a security system in place that records everything and even alerts you when something has happened to the environment, you can now have evidence. Furthermore, if you decide to have environmental sensors in the house, you can detect moisture problems and other related items very early on.

AC and Maintenance Monitoring Keeps Value

Security is a huge deal when it comes to residential houses and apartments. Many people like the idea that they live in a very secure place, which can easily be provided. However, you can also add on top of that security to lower the cost of repairs. Security footage will catch damage that has occurred to the house. Humidity sensors and oxygen filters with sensors inside of it will detect gas problems, water problems, and air quality.

Oxygen filters with sensors in them that allow for the detection of gas problems prevent fires. Air purity sensors will detect humidity and compared to the outside humidity to determine if the house is unusually human. Often, when a water pipe breaks inside the house the extra water builds up heat and this shows itself in the form of humidity. Not only that but you could add a sensor that detects the amount of pressure running through the water pipes in the house. Air purity would help you detect when tenants are smoking when they're not supposed to be because that would show up on an air quality report.

By having these sensors all over the place and having cameras, what you do is you have ecosystem that lowers cost. Yes, it is expensive to invest this into a house but if you are trying to rent the system out or you're trying to sell the system out, it adds monetary value. It also ensures that your house does not see a massive amount of damages outside of abnormal weather conditions. It protects your property from theft, vandalism, and even arsenic in some cases.

AI IN MARKETING OF REAL ESTATE PROPERTY

How AI can be used to target ideal clients. Make a mention of some of the methods used in the previous books; and link to the book.

Email Targeting

Email Lists Are a Thing of the Past: Chatbots

If you are still doing an email list, you are likely hitting the oldest Generations there are. Ironically, the way of the email list is going away. Instead, what you have now is the instant messenger list. This list is on Twitter, Facebook, Instagram, and pretty much any social media website you might think of. This is because most of the younger generation deal with instant messenger and email is not really a priority on their list. That being said, professionals still use the email list but you're going to reach a far lesser audience.

Therefore, while it isn't ideal to constantly focus your attention on the email list, it's a good idea to probably keep them around. The reason why the shift has been going over to the instant messenger is that it opens the possibility of a chatbot. Essentially, people use the instant messenger to sign up to your chatbot list. Every time you have a real estate listing, you simply blast this out to everyone in the list. Then what happens is that the interested party chats back.

This is where new clients or potential clients encounter artificial intelligence. The chatbot can respond to them and ask them for information. It would be very similar to the chatbot system you would have to pad out the details you might need to sell them a house. Therefore, if they are interested in buying a house the common facts can be listed and if the client is still interested, you can directly connect to them via instant messenger. This means the client can go from being engaged with the artificial intelligence to being engaged with you on the selling end.

Always Keeping Clients Engaged

The purpose of this chatbot is to keep a constant engagement with your client. And an email list, your user is very likely to just sit through hundreds of emails a day. On a constant basis, even I receive about 100 emails from every which way under Sunday. There's only so much you can interact with via email and it is quickly overloaded by password changes, commercial website deals, and generally the bulk of email is buying stuff from websites.

With a chatbot, since you're selling singular items like a house, you're not going to be messaging them constantly. Instead, if you collect an income level you can assign them to a specific bracket of your broadcast. However, there's way too much to cover in this book so if you want to learn more about that you can actually check out this book here.

Average User Willing Purchase Range

By being able to put them inside of this bracket, you effectively create your average willing price. People who are actively searching for houses that give you their income level can also have an asking price. This asking price can determine what houses you should be really looking to sell to your clients. Rarely do people want to buy $300,000 to $500,000 houses. The most common range to sell within is between $90,000 houses and $250,000 houses. This is because much of the United States population is either middle class or poor. Very few people live in the high enough bracket to afford a half a million-dollar house.

By creating this bracket and determining your average price, you don't accidentally buy houses that rarely sell. Instead, you begin cycling through houses every week because you're constantly selling them. In fact, the benefit of a chatbot is that it is so easy to spread the word of mouth. You can have them share a post that says they bought a house from your chatbot. Since someone reliable participated in your chatbot and shared it with their friends, those friends know you have your friends. So long as you keep selling houses to satisfy customers, the word of mouth advertising becomes instantaneous.

Finding A Group

Click funnels

A click funnel is a concept that's been around for quite a while, but not the artificially intelligent click funnel. These funnels take information about how successful your attempts have been to generate customers and build a profile of what will likely work. Therefore, say you built five pages where only two of the pages have done well. You can then run Artificial Intelligence on this to determine how successful a new page would be and what should go in that new page.

However, engaging consumers is a huge topic and there's simply not enough room to talk about how you can use artificial intelligence to do market strategy engagement. I do have a book on market strategy engagement that does cover click funnels as well as other artificially intelligent marketing strategies that will keep your customers engaged and you can find more about it here.

Real Estate Affiliates

There is one strategy that I don't talk about inside of the book however and that's because this only applies to certain use cases. You see, you really want customer engagement in real estate with a single customer. This means that the user will come to your website or service to engage with you to buy a house. Once that process is done, you are not really interested in constantly engaging them. You want to entice them, and you want them to hear your deals so that they will buy them, but you don't need to retain customers after you've already sold the house. This is where affiliate sourcing comes in.

We often hear about affiliate links whenever we are dealing with YouTube videos or even standard blog websites. It's another avenue to make money for these creators. What is rarely talked about is the real estate affiliate market because it's rather small, but it does exist. If you are interested in selling run-down homes that you get off the auction, this could be an option for you. If you are looking for cheap contract labor that's always moving, this may also be a market for you. It really depends on how you use it. Essentially, you would just offer to pay a percentage of the commission you get for selling a house. For each house sold, the YouTuber would get an amount of money to justify putting your affiliate link in their description. Since YouTubers are currently undervalued at this point and they reach millions of people, it's cheaper to do this than to go to an actual advertising firm or even a Blog website.

Getting in The Real Estate E-Circle

The true goal here is to get into the Real Estate Circle in the online world. This is more of an abstract concept rather than an official thing. The Real Estate Circle refers to a gathering of people using the same tools to sell multiple houses at a time. It's very similar to a company but you can utilize the same platform to sell houses. Instead of a company where everyone gets assigned houses or assigned specific tasks, the marketing and engagement is done by you with shared tools. This prevents other real estate agents from taking your clients, but it also allows you to share clients depending on what houses you're selling.

Let's say that you're having a hard time selling a house that's worth $120,000. Of the customers you have been engaging, they are looking for houses within the ranges of $140,000 to $180,000. For weeks on end, you have not engaged a customer looking for that low of a price range. However, because you use a platform where you share tools with others and clients with others, one of the other real estate agents who might have found someone like that can share that client with you. This allows you to sell the house while still focusing on the proper range that you are in and not messing up that range from refocusing your efforts. This allows for better sales figures and faster sales figures.

AI IN FINANCING OF PROPERTIES

Introduction of Artificial Intelligence in the financing of properties is an exciting prospect. It involves ensuring that the most suitable applicants for the property are seen by the landlord. While credit score and current income have been used traditionally to gauge the ability of an applicant to pay the landlord; they ignore a lot of fundamental issues such as an applicant's debt, credit limit, future job prospects and situational awareness. This is an exciting development as it ensures landlords are able to get the clients that are most likely to pay back; while it doesn't impose unnecessary debts on anyone who is unable to pay them back.

Personal Path of Growth

Educational Growth Determines Revenue

This is an easy measurement to set for your artificial intelligence. Nearly every successful person boasts about their education. Therefore, you simply must run a test on successful

139

people and educational status. This will create the artificial intelligence you need to judge a person by their education.

Most billionaires and millionaires will state one common thing. For argument's sake, let's say that one common thing is that they read a book every day. Therefore, while they don't always necessarily conform to standard educational practices, like going to college, they do learn constantly. Therefore, you can base a person's success in the overall life by their educational patterns.

Learners Are Self Reliant

The important aspect here that you want to focus your Artificial Intelligence on is how much knowledge is institutional or self-taught. Institutional knowledge shows a willingness to proceed to success. Self-taught knowledge shows a willingness to proceed to success in the face of adversity. In other words, the self-taught individual has a higher commercial and ambition-based drive than the other person. In addition to this, it also means they're more resourceful than the other person.

To be a self-learner, they had to specifically look for the resources. They had to put in the effort to find the right material. In addition to this, they likely hit a lot more walls than the institutional learner did. This is because self-taught learners have a habit of making more mistakes and being rejected more, which means they fail fast. Another common thing that millionaires and billionaires share is that they often tell you to fail fast.

DIY Environment Promotion

The do-it-yourself community has grown into a giant behemoth these days. It is very difficult to go every day without seeing some sort of do-it-yourself video. You might think that this is rare for you but think of how many times you look up how to do something on YouTube. This is how the do-it-yourself community works.

This one is a little bit harder to do and is a bit subjective. The way I would go about doing this is by simply counting how many items there are in the "how many things do you know that are DIY?" as this provides me with a simple numeric value. This would give me a score and I could then break up that score into an area map. The area map would put an individualistic score on associated words to related jobs.

This would tell me how many people did their own makeup versus how many people know how to fix a water heater. An area map is a collection of terms gathered in an area in the visualization of a circle. These terms are related to a specific keyword that you used. They are an easy way to see where a person is very dominant in relation to their activities.

I kind of gave you a hint but the reason why this would be useful. Sometimes they might just fix the problem themselves. If you can get a person who understands a lot of DIY about houses, they might fix it before they ever call someone. This means that they'll spend less money and be more profitable overall.

Skills Reliability

You can then take the information that you understand about their education and their DIY skills and do a skills reliability test. A skills reliability test is simply finding the most related job field that their skills are associated with and seeing how many people hold that position. The more people in a specific position, the more likely it is that that person can be replaced. This is because more people know this topic and so labor is more easily found. You don't want it too low because it might be a very niche type of work. Therefore, you must find the right threshold, something in artificial intelligence would do.

By doing a skills reliability test, you determine quite a few things. The first thing you determine is the likelihood of how high their income is going to be. The second thing you determine is how hard it is for them to be able to find a job should they be fired. The third thing you determine is if there is a progressive path in their field of skills.

Understanding how high their income is going to be will help you determine how high your financing rate can go in the future. Determining how difficult it would be for them to find a job given their skills determines whether they are going to default more likely or not. Lastly, understanding if there's a path for progression in their field allows you to know if they are pursuing a "learning" based field. As I've mentioned before, learners will always tend to climb the financial ladder.

Debt

Paid Vs Unpaid

It is very important to understand if a client is willing to pay off their debt. The number one thing you don't want to see as a person who finances is a customer who never seems to pay off their debt. A credit report will tell you how many accounts they have open as well as how full those accounts are. The credit score is a bit ambiguous, but he usually represents how often they pay off their credit cards.

Therefore, the first step is almost always to check whether the customer usually pays off their debt. Additionally, there is something unique about certain debtors. Some of them will completely max out but never be late on a payment. They will continue to max out, but they will keep up with payments. This is a debtor you do not want to have.

Revenue Vs Credit Line

This is where understanding their overall revenue versus the amount of credit line they have is important. Really, a person should only have enough credit line to take up half of their annual paycheck. If they have more than that, they run into a situation where no matter how much they work, it will take years for them to pay off their debt.

In such a situation, if you finance them, you run into the situation where they prioritize debt. Yours will likely be the debt that they prioritize but some of their debt is like Care Credit. Care Credit is a service where the credit line is specifically dedicated to those who need help paying medical bills. This means that they could easily rack this credit card up without even meaning to. If they get annual checkups like most people are supposed to, this credit card will have around $500 on it every year at the minimum. If they go into the emergency room even once, that balance suddenly shoots up to $2,000. I'm not saying that all emergency room visits will be that price, but they generally tend to be within that range. I remember a couple of years ago being charged $500 for just entering the emergency room.

Essentially, if they have more credit line than they do revenue, they are in a situation where they max out their credit then there's no more revenue to offset the cost. Let's say that our potential tenant earns about $24,000 annually. That tenant has $24,000 of a credit line. Most cards will have either a $40 minimum or a 2% minimum. This means that if they max out their credit line then the minimum they would pay is $480 for the entire year. If they are a customer of yours and you charge them $1,000 monthly, they have plenty of money to spend towards paying off that credit card and then paying off the finance. This is because it's one credit card. However, interest plays a funny role in this.

Let us say that the interest rate on this credit line is 5%, a very nice healthy 5% (not realistic). That's an extra $100 a month on top of $40 a month. Now, it doesn't matter if you split that amongst multiple cards. However, if the cardholder only ever pays the bare minimum then that's really $60 a month. That means they will pay their card off in 400 months versus 171. Thus 33 years versus 17 years of debt.

Deferred interest is the nasty bug here, which Care Credit has. Deferred interest is more likely to be 30% of the original balance the deferred payment is on. So, let's say something horrible happened and The Deferred interest is on the $24,000. A person who's going to take 33 years to pay of the regular balance is no match for paying off $24,000 in a single year let alone $31,200 when that interest hits. Their annual bill goes up to $624 with interest hitting $1,560. If they're making $24,000 then they will never get out of debt.

Before we hit Deferred debt, the client had a possibility of paying off the debt. Once the special clause of Care Credit hit, the client no longer had the ability to pay off their debt. This would lead to bankruptcy and you would be outed the money you financed. Therefore, paying attention to the *types* of credit line they have is very important. It's a hassle to do this every time, but, with AI, you can predefine rates. You can skip this hassle and the AI will just tell you the risk of the client you're looking at.

Intended Future Investments

Another trick that you can do that is very similar to a personality test is determined by their future Investments. Let's say that you might be financing a college student. A future investment would be the type of job that the college student is going to college for. Therefore, you can have artificial intelligence investigate the success rate of that job and degree. So, if a college student is going in for computer science and they are looking to become a web developer, you can look at the hire rate for web developers. More importantly, you can look at the hire rate of that job for the area of the house you're financing.

Since its web development, odds are high that they are likely to find a job. This means that they will get higher pay and they will be able to pay off what you finance them for much faster. On the other hand, if they took a course for gender studies, this would be a different matter. Gender studies do not have a fantastic job score when it comes to a hire rate. While the current generation of college students may think that such a job is important, there aren't a lot of job roles those people can fill. This means they will often find themselves jobless. This also means that they are a risky investment. However, this kind of has to be handled by artificial intelligence.

Job rates change a lot. For instance, with web development, front end development was exploding nearly two years ago. Now, you'll find that most companies that look for front end development are in the bigger cities. This is because a lot of people rushed to fulfill this role. Likewise, the job rate fell significantly. This is just one industry with one specific type of job. If you think of all the industries that are out there and their specific jobs, you begin to understand why you need artificial intelligence. A single human simply can't go through all those different jobs and keep a consistent job rate available for comparison. Instead, a single human would do it on a per basis average. Therefore, time would be wasted every time a client sought out funding. Therefore, artificial intelligence, which can do it all the time and much faster than humans, is perfect for this type of job.

Offering Paths Bait

Likewise, you can offer help to go through different job industries. If a person is seeking finance, they usually do so because they cannot handle the upfront cost. This means that their job doesn't adequately support what they want to finance for. Knowing this, many companies still financed clients. This is because usually, the financing is just to get over the hump of getting into a place. Once they are in that place, they tend to do very well. The problem comes when you want them to pay that financing off faster.

You always want ambitious people to be the target for your financing. An ambitious person will try to get the highest paid job that they can when they can. Therefore, if you have an artificial intelligence design to find the most profitable jobs, you can then entice them for a better future. If a person denies this type of offer, it means that they are more comfortable where they are and not ambitious. That means that if they are fired, they are not likely going to be able to adapt to another job quickly. That means that should something happen to their source of pay, they are very likely to not be able to finish paying off what they owe. I've known several people who can pay their bills by working at

McDonald's, constantly complain about how they don't ever have any money, and yet I see tons of job opportunities that they could fill that they don't bother. Years later, while I write this book I still see them at their same jobs. They still owe money and they still complain yet their pay level hasn't really changed. They are essentially stuck in that position because they are comfortable in it. You want to avoid these types of people because these types of people, when they lose their job, they're not going to be able to easily transition into another. This is because they don't bother learning new skills as a result of being comfortable.

Situational Awareness

Disability Compensation

Situational awareness is important when you're considering financing someone. Sometimes, you run into some unique scenarios that benefit you and your customer. Someone is looking to purchase real estate, but they don't quite meet the qualifications yet. Essentially, they don't exactly have the money to do it. However, they do have a unique ability that sets them apart from others.

For instance, if a person is disabled then they can get money for being disabled. Normally, the person who is disabled will get a monthly paycheck from the Social Security office. This represents a stable income that is not likely to go away. In some instances, the type of job that this person does could be classified as a hobby. You can make money from a hobby and collect disability in certain situations.

You do get taxed a lot more for hobbies than your regular line of work. However, a hobby can be classified as non-reliant income. Therefore, someone may make near $24,000 annually from a hobby and still collect disability. It's a very interesting loophole that very few people take advantage of and even fewer people I have available to them. However, if you have artificial intelligence looking for these types of people, then you can take advantage of that. Since the rules are laid out by tax law, that means you have a definition of the type of person that would fit that scenario.

Essentially, tax law defines what machine learning you need to take advantage of this type of person. Now, it's important to also understand that taking on this type of person might mean they die and you don't get paid. However, if you are looking to invest into a retired couple, they are very likely to stay healthy. In this situation, you might even be able to take advantage of owner financing.

First Home Bonus Help

You could also use artificial intelligence to find first home buyers. In the United States, many states allow the first home incentive. This incentive was utilized to increase the number of young adults buying home. In other words, it was meant to increase the amount of tax revenue the state could claim. However, many young adults don't even know this law exists. In fact, many young adults rent for the better part of their lives. This is because they tend to not know anything else and don't realize how cheap owning a house can be.

However, just because you have the first home incentive, doesn't mean every young adult is eligible to it. There are requirements that a young adult must meet in order to purchase a home using the first home incentive. This is a repetitive task that can be handled by artificial intelligence. Since there are already rules in place, it defines what you need the machine learning algorithm to detect.

Incentivizing Home Improvement

Sometimes you have clients that can't pay the average price. Let's say that you have a family that is on state benefits. In other words, combined they make about $3,000 a month. This would be about $1,500 a month for each of them. This would qualify them in most states as being below the poverty level. Now, let's also say that you have a home that isn't in the best of shapes. If the family is trying to get the lowest monthly bill you

155

can use artificial intelligence to detect that. Odds are is they are trying to go after low-end homes. Artificial intelligence might ask them what their income is, how much they're looking to pay and if they are on assistance.

Contractors can be rather expensive, but you don't need contractors unless you are doing specialized work. If you're doing anything electrical, you need a contractor. However, if you are just repairing what's there then it can be handled by a person willing to watch YouTube. Therefore, artificial intelligence can begin asking about skills that they might have. Therefore, if they have an understanding of how to put down tile and there's a lot of cracked tiles, that's a bonus. As you can see, this could go for nearly every skill. This provides you with a situation where you don't charge them as much for financing in exchange for improving the house. This would be ideal for sellers who just acquired a house and people are looking to buy it. You don't have to give up your house right away because they'll enter a contract where you owner finance them, but if they ever default or become unable to pay then you then you can re-obtain the house with improvements.

You might do this for a few reasons. It's too costly for you to do the repair, they are likely going to end up giving the house back, or you haven't been able to sell the house for a very lengthy time.

CONCLUSION

AI Replaces Repetitive Tasks

Artificial intelligence can be used for several things, but it's primarily used for repetitive tasks. Things like filing documents, doing taxes, finding homes, researching neighborhoods, and all of this can be done by artificial intelligence. When you save time on things that don't make money, it makes you eligible to use that time on things that do make money.

However, that isn't all that artificial intelligence can do. With artificial intelligence, you can have a customer support channel that you might have not had beforehand. This customer support channel would help save you time by answering customer questions.

It is rather easy to build artificial intelligence because artificial intelligence is simple. Artificial intelligence is a combination of binary choices. In other words, if you can break a problem down into yes or no answers, you can use artificial intelligence on it. This means it can handle a lot of the work.

However, that doesn't mean it can handle all the work. For instance, writing this book cannot be done by artificial intelligence. The reason it can't be done is that of narrative. You might have heard of an artificial intelligence that was sold as an article rewriter. Specifically, companies like BuzzFeed produced artificial intelligence to write articles. Rather, they used artificial intelligence to find trending articles.

If you use artificial intelligence to talk to other humans, it won't be able to understand context. Narrative drives a story, it goes from A to B and therefore it needs context. The way that humans learn is that we learn the first item and then build on the first item we learn. The way a neural network learns is by making a combinatorial pattern of decisions that are optimized. Therefore, humans and neural networks learn very differently.

There is some similarity to humans and neural networks because to expand our knowledge, we must make decisions. In fact, a combinatorial pattern of decisions is how we humans' practice what we learn. Rules are defined ahead of time, we make decisions based off those rules, and then someone else corrects us. This is very similar to how a neural network works but we have a linguistic advantage.

The linguistic advantage allows us to instantly translate input into context and vice versa. Artificial intelligence needs humans to translate context into input and output into context. Until artificial intelligence can translate context on its own and turn the output into context, it cannot do a job like this. Our language, by default, is a narrative language. It is a descriptive language whereas the machine learning language is a numerical language.

This is the primary reason why humans have advanced so quickly, we can create associations by default. Machines must be told what associations should be made. We tell these machines how the associations can be made with mathematics. Sometimes the mathematics is simple, sometimes the mathematics is on the level of quantum physics. However, in all cases, humans must translate context from a linguistic standpoint to a numerical standpoint.

This has its downsides and its upside. On the upside, so long as we can translate the context into numerical language, machine learning can do the job for us. The downside is that there are some things that machines will simply never understand. This means that there are some jobs that are truly irreplaceable by artificial intelligence. Specifically, the job that requires invention is the most impervious to artificial intelligence.

It was thought long ago that if we could create artificial intelligence, that artificial intelligence would advance on it. In other words, artificial intelligence could create new mathematics. The problem is that mathematics almost always comes from curiosity. The curiosity about light, gravity, physics, and the whole lot use a contextual event that is not numerical. Yes, artificial intelligence can certainly become very advanced. However, it cannot pursue curiosity. We have yet to define curiosity into numerical terms. Once we manage to define curiosity into the numerical language, that is when General AI will exist. That is when all jobs will be replaceable.

Security Bonus with AI

Artificial intelligence is very good for handling multiple things at once. So long as you can define what is supposed to happen and what is supposed to not happen, you can generally secure a place very easily with artificial intelligence.

The first element of security that you can ensure with artificial intelligence is property damage. Whether it's caused by the tenant, an outside force, or a problem in the utilities, artificial intelligence can detect it. Not only can artificial intelligence detect an issue has occurred, but it can identify what's occurring and let the owners know. Therefore, if a tenant has broken something then artificial intelligence can identify an oddity. If someone has broken into the house, both the tenant and the owner can be notified.

This allows for the opportunity for the tenant to inform the owner not to call the cops. The tenant or the owner can call the cops if they need to. The alarm does not need to go off for the entire house while the cops are on their way. In fact, one could say that an alarm goes off in the bedroom because of something like Amazon Echo. Lastly, utilities inside of a house can have The Internet of Things sensors on them to prevent problems from cascading. You can detect abnormal moisture, increased or lowered pressure thresholds, and even the amount of times the house needs to be cleaned. This will prevent water damage and potential bug infested homes.

The second element of security is holding people accountable for what has taken place. Maintenance workers are often accused of stealing out of people's homes. However, it's very hard to prove it unless the place they are living in has cameras. Even then, it doesn't always provide one with the ability to determine who that maintenance worker is on the video footage. Artificial intelligence can associate the key of the maintenance worker, whether it be an RFID chip or facial recognition, with the video footage. This means that the incident can be recorded to match the time slot.

The average method for determining whether a maintenance worker stole from a resident is tedious. First, the resident places a complaint and then potentially calls the cops. Then, the security team goes back through footage over the entire, potentially, week to find out if anyone entered that residency. This means that the resident must wait a ridiculous amount of time before they find the criminal. By then, the criminal would have likely sold whatever they stole. The only evidence to say they ever took anything would be video. With artificial intelligence, the time the maintenance worker entered and left could be recorded. Artificial intelligence would segregate video footage by room. Then security would look up the room it was stolen from and maybe review an hour and a half of footage. I say an hour and a half given 30 minutes for each cleaning session three times a week. By the time they were done, security would be catching the criminal an hour and a half after the resident found out that something was stolen. In that time frame, it would be very difficult for the criminal to sell the item.

Another weird but very doable circumstance is to make sure people are doing their jobs. The most common way of a workplace keeping track of whether a person is doing their job or not is a checklist on a board. People fake this all the time, but artificial intelligence can do something more. You see, if a maintenance person uses their key then artificial intelligence can begin assessing that person's job. The artificial intelligence has memorized the motions needed to do specific jobs like dusting or sweeping. This way, the artificial intelligence might checklist sweeping off from the kitchen camera. Furthermore, the kitchen may be attached to the dining room but the difference in flooring allows the artificial intelligence to segregate it into different rooms. This would be far more effective than a checklist. It would also make sure that the job is done right and footage is recorded. This could prove the maintenance worker wrong or catch when the artificial intelligence makes a mistake.

The third element is in the form of an advertisement. People like knowing that they live in a secure neighborhood and a secure house. In fact, some people need security for their jobs so even though they rent, they wind up putting a security system in place. Having additional security with the backing of artificial intelligence becomes a huge selling point for real estate. It can justify a doubling of a price tag for rent or increasing the real estate value by a third of the price.

If you take everything I just mentioned and included with a rental or a community home, it serves as its own advertisement. Knowing that people's stuff is safe while they are out and knowing that it's difficult to get in the home, people will be enticed by these two facts alone. If a person buys a 70-inch television, they do not want to think it will be stolen. if a person buys the newest $2,000 gaming rig, they don't want to think that a maintenance worker will walk in and lift it. Understanding this allows you to advertise to people with bigger pockets that are willing to spend a lot of money. The ironic part is that such an investment is usually not that very high. The cameras may cost somewhere around $50 to $100 a piece. The computer that controls the artificial intelligence May wind up being a $10,000 investment. However, if you live in a 50-apartment complex and you use a $1000 security fee, you're looking at $5,000 a month. In a year, because you probably made a $30,000 investment for security, you're making nearly $60,000 a year. It's a relatively low investment that benefits you significantly and justifies a higher price.

Better Pricing with AI

You also have the added benefit of better pricing with artificial intelligence. Not only that, but it's better pricing with buying and selling. This is because artificial intelligence can be used in the same way with slight differences.

When buying a new house, you tend to pay attention to quite a few details. These details come in the form of a mental checklist of items you prefer over items you detest. Therefore, if you are the type of buyer that looks for homes that are run down but only have cosmetic issues, you might detest a listing that says water needs to be installed. In some cities, real estate is often required to have water pipes leading to city water. This is forced on the residents of the city and usually cost about $20,000 to $40,000. this is a cost you basically must absorb up front until you can sell the house, which is sometimes not ideal. So, while you may look for houses that are in a worse for wear shape, that does not necessarily mean that you want a gut the house apart.

On the other hand, artificial intelligence can help when it comes to selling. You look for the same things about buying a house that you do selling a house. You look for the quality of the school. You look for crime statistics. You look for the average revenue. You look for the actual cost of the materials. These are all standard things that you look for whenever you are trying to buy a house or sell a house. In terms of buying a house, you want to look for these things because it means that house is likely to be worth more. Therefore, if you can get a house for a good price in a good neighborhood with low crime, you can turn it over for a lot of money. Likewise, when you go to sell a house these are the things that you advertise.

Now, you may be new to the real estate industry and you may be looking at artificial intelligence to make more money. If this is true, you're likely wondering why I say you look for these things when you're selling. When you fix up a house, it takes time for that house to be fixed. Usually, if you're going to find a house that's about a $90,000 buy in a neighborhood of houses where the average house is over $200,000, you're going to spend a lot of time there. Within a single year, you can see a high-quality school become a medium-quality school. You can also see crime statistics go up if a new law is passed or if a certain commercial market came into place. Likewise, those in the neighborhood might have moved out of the neighborhood. Therefore, it makes sense to do a repeat of the research you did before to get an actual estimated value of the house that you are trying to sell.

Since the variables are the same but the outcome is a different intention, artificial intelligence can be made to handle part of it. Artificial intelligence can be made to judge a house's value based on the factors you investigate. If you are buying a house, you can have the output tell you whether the house is worth buying. If you are selling a house, you can have the output tell you whether the price you want to sell it for is reasonable. Since artificial intelligence is programming, you don't have to have two of the same process to get two different results. Instead, you can have the same process but, at the end or near the end of the code, you can set up the different cases for buying and selling. This is because programming is modular and components can be reused.

Embrace The AI

Artificial intelligence is not new but the way it's being used is new. You might wonder whether it's going to replace you as a real estate agent. You might even wonder about whether buying and selling houses will become artificial intelligence only. The truth of the matter is that you can't escape it. Even if you don't want to use artificial intelligence, your competitor real estate agents will. This means that, ultimately, all you can do is join the side that uses artificial intelligence.

Artificial intelligence is not that scary. It's simply a matter of yes or no. If a house is in an area with two schools, here's how the artificial intelligence is going to look at it. It's going to run a search algorithm that determines how many schools are within a certain mile radius. It's going to ask what letter grade the school has been given by the state. If that letter grade is above a preset value, it then passes on a yes state to the next variable. While that neural node does that, another neural node is looking at another school and either sends a yes or no. What the neural node sends is usually a one for yes and a zero for no. If the next neural node in the chain after those two neural nodes is receiving a yes and a no, the ultimate result is a 0.5. If you combine that with however many schools are in the area, you might end up with a 0.3 or a 0.6 or even a 0.9. Ultimately, it looks through and just adds in the scores together to give you an average overall score. This overall score determines the quality of schools in that area.

As you can tell, it's not really that difficult and that's because what you're looking at is not that complex. If you're looking to buy a house, you look at crime statistics and school quality or whether something exists in an area. School quality is often determined by letter grades and crime statistics are already numbered. Therefore, all you do as a real estate agent is look at the number to see how high it is for crime statistics. In fact, you do the same thing with schools. The only difference is that with schools, you're looking for a higher number. Meanwhile, with crime statistics, you're looking for a lower number. That means the average calculation that you must do is simplistic and repetitive, which means that artificial intelligence can do it. As you can tell, it's not like you're dealing with a thinking, breathing machine. You're just dealing with a very fancy calculator and there's honestly no reason to fear a calculator. Over time, we might create something known as general AI and that is something to be wary about, but not a calculator. So, you can either embrace artificial intelligence and make it part of your workflow or you can retire when it's part of everybody else's workflow. The choice is up to you.

www.ingramcontent.com/pod-product-compliance
Lightning Source LLC
LaVergne TN
LVHW022318060326
832902LV00020B/3539